Wissenschaftliche Beiträge
aus dem Tectum Verlag

Reihe Nachhaltigkeitswissenschaft

Wissenschaftliche Beiträge
aus dem Tectum Verlag

Reihe Nachhaltigkeitswissenschaft
Band 3

Viktoria Drabe

Innovating in a Circular Economy

Exploring the Case of Cradle to Cradle Implementation

Tectum Verlag

Zugl.: Dissertation an der Technischen Universität Hamburg, 2021

Originaltitel: Exploring why and how companies succeed in the shift towards a Circular Economy –
The case of Cradle to Cradle innovations

Die Deutsche Nationalbibliothek verzeichnet diese Publikation in der Deutschen Nationalbibliografie;
detaillierte bibliografische Daten sind im Internet über http://dnb.d-nb.de abrufbar.
The Deutsche Nationalbibliothek lists this publication in the Deutsche Nationalbibliografie;
detailed bibliographic data are available on the Internet at http://dnb.d-nb.de
ISBN 978-3-8288-4742-2 (Print)
978-3-8288-7842-6 (ePDF)
ISSN 2749-0092

British Library Cataloguing-in-Publication Data
A catalogue record for this book is available from the British Library.
ISBN 978-3-8288-4742-2 (Print)
978-3-8288-7842-6 (ePDF)
ISSN 2749-0092

Library of Congress Cataloging-in-Publication Data
Viktoria Drabe
Innovating in a Circular Economy
Exploring the Case of Cradle to Cradle Implementation
Wissenschaftliche Beiträge aus dem Tectum Verlag: Nachhaltigkeitswissenschaft, Bd. 3
192 pp.
Includes bibliographic references.
ISBN 978-3-8288-4742-2 (Print)
978-3-8288-7842-6 (ePDF)
ISSN 2749-0092

1. Auflage 2022
© Nomos Verlagsgesellschaft, Baden-Baden 2022
Gesamtverantwortung für Druck und Herstellung bei der Nomos Verlagsgesellschaft mbH & Co. KG
Printed in Germany

Alle Rechte, auch die des Nachdrucks von Auszügen, der fotomechanischen Wiedergabe
und der Übersetzung, vorbehalten

This work is subject to copyright. All rights reserved. No part of this publication may be reproduced or
transmitted in any form or by any means, electronic or mechanical, including photocopying, recording,
or any information storage or retrieval system, without prior permission in writing from the publishers.
Under § 54 of the German Copyright Law where copies are made for other than private use a fee is
payable to "Verwertungsgesellschaft Wort", Munich.

No responsibility for loss caused to any individual or organisation acting on or refraining from action
as a result of the material in this publication can be accepted by Nomos or the author.

Foreword

The understanding that natural resources are a finite good and should be used with prudence is becoming more and more commonly accepted in society. With regard to our current economic system, however, it also becomes clear that this awareness is often difficult to realize in practice, as the world's resources continue to be constantly depleted and internationally agreed climate targets are repeatedly missed. At the same time, our economy, defined by a linear direction of material flows and characterized by a take-make-waste approach, is increasingly confronted with a dramatic depletion of finite resources, an increase in price volatility, and a growing customer expectation for product sustainability.

Understandably, this outlined understanding repeatedly spurs the discussion and search for concepts that open up opportunities for the environment and the economy alike and are intended to resolve the seemingly inherent conflict between economy and ecology. The idea of a Circular Economy (CE), as promoted by the Ellen MacArthur Foundation and its associated companies and stakeholders, has been the subject of intense debate in recent years and represents a particularly promising concept. CE describes an alternative paradigm representing an industrial system that is restorative and regenerative by intention and design.

In this work, Mrs. Drabe examines the motivations of companies embarking on the Circular Economy journey and how they can be successful in doing so. For this purpose, she focuses on the Cradle to Cradle paradigm (C2C) with its associated certification program, which represents the state-of-the-art to transfer the idea of the Circular Economy to the operational implementation level. Her empirical field is formed by companies that have implemented this program and have gained corresponding experience in implementation and certification.

While the research landscape so far contains numerous studies on the definition, delineation and conceptual implications of alternative concepts of an environmentally friendly economy, there is little empirical evidence on the enablers and barriers of actual, operational implementation. To the best of my knowledge, the present dissertation by Mrs. Drabe is the first comprehensive, empirical study on the implementation status of Circular Economy in industry using C2C innovation as an example. In this sense, Mrs. Drabe is doing real pioneering work and has presented what I consider to be important and insightful work.

The results of the qualitative interview series not only provide initial, interesting findings on the problems, challenges, and opportunities of C2C implementation and certification, but also allow her to develop informed propositions that in turn form the basis for the quantitative exploration she subsequently conducts and lays out by means of an online survey. She develops the underlying variables from existing related theories and through newly developed constructs from her qualitative study. Following an exploratory factor analysis, she elaborates on the impact of critical motivational and organizational contextual factors with respect to the organizations' satisfaction with the implementation. Based on the extensive and insightful discussion of her research findings, Mrs. Drabe elaborates two frameworks to be able to build an adequate organizational environment for a substantial embedding of CE at the organizational level.

Overall, this work makes a distinct contributions to Circular Economy research and provides valuable recommendations for companies regarding the design and marketing of C2C products and services as well as the related internal process changes. The quality of the research results combined with the very knowledgeable application of scientific methods as well as the insightful interpretation and precise presentation of the results validate the research approach chosen by Ms. Drabe. For me, the essential contribution of the work lies in the very well-founded theory discussion and expansion as well as reappraisal of a very actual as well as significant phenomenon. In this respect, Mrs. Drabe makes an important contribution to science as well as to decision-makers in corporate practice and politics.

Hamburg, December 2021   Univ. Prof. Dr. Dr. h. c. Cornelius Herstatt

Acknowledgements

The current economic system, mainly building on a linear direction of material flows, is increasingly being confronted with a dramatic depletion of finite resources, increase of price volatility and a growing customer expectation with respect to product sustainability and quality. The concept of Circular Economy (CE) describes an alternative paradigm, especially with respect to innovation and product design.

Decoupling growth from resource use, CE provides a substantial business imperative, which an increasing number of companies and governments start to recognize. From the research perspectives, the research landscape hitherto covered numerous studies on definitions and conceptual implications, while there is little empirical evidence on enablers and barriers of actual CE implementation. This points at a research opportunity with a focus on organizational implications of CE. Given the nascent stage of research, the work uses a hybrid approach consisting of a qualitative and quantitative study to find answers to the central research questions of why companies decide to engage in CE efforts and how the organizational context affects its implementation. For this purpose, the Cradle to Cradle (C2C) certification, a closely related standard which is granted to companies based on specific assessment criteria, was selected as the empirical field.

Based on an exploratory factor analysis and a subsequent multiple linear regression, the work lays open which parameters can foster an adequate organizational environment to substantially anchor CE on the organizational level. The conclusion provides valuable guidance not only from the research perspective but also includes managerial and political implications in order to spur the transition towards a Circular Economy.

Throughout the writing of this work I have received great support and assistance for which I am very grateful. First, I would like to express my sincere gratitude to my interview partners and survey participants who not only made it possible to accomplish my research goal by providing the most valuable input, but also encouraged

Acknowledgements

my endeavour by cheering the purpose of my research project and underlining its relevance for them.

I would also like to thank my academic supervisors, Prof. Dr. Dr. h.c. Cornelius Herstatt and Dr. Armand Smits for their constant support, guidance and inspiration in many constructive discussions. Your feedback helped me to sharpen my thinking and brought my work to a higher level. I would also like to acknowledge Prof. Dr. Dr. h.c. Wolfgang Kersten for chairing the doctoral defense in such a professional and at the same time solicitous manner.

Furthermore, I feel privileged to have met and worked with an amazing set of colleagues, many of whom became friends. You have contributed ideas, were always open for discussions and made the PhD time a truly fun and memorable experience.

I would also like to thank my family and friends for their ongoing support and encouraging words, their patience and happy distractions during harder times. In particular, I would like to thank my parents and my sister for their endless love and believing in me.

Finally, David, my husband and father of our two girls, I need to say no more than I could never have done this without you. Thank you.

Hamburg, January 2022   Viktoria Drabe

Table of Contents

Foreword	V
Acknowledgements	VII
Index of Figures	XI
Index of Tables	XIII
List of Abbreviations	XV

1 Introduction 1
1.1 Research relevance 2
1.2 Research objectives and contribution 5
1.3 Structure of the thesis 6

2 Phenomenological background 9
2.1 Brief perspective on the historical development of sustainable innovations 9
2.2 Evolvement of the Circular Economy concept and underlying schools of thought 13
    2.2.1 Main principles and core ideas 15
    2.2.2 Critical discussion of the CE concept 17
    2.2.3 The CE as an imperative for future business 18
2.3 Selection of C2C as empirical field 20
    2.3.1 Origin and development of C2C 20
    2.3.2 Main characteristics of C2C 23
    2.3.3 C2C on the organizational level: the certification program 26
    2.3.4 Challenges of C2C implementation 29
2.4 Interim summary and derivation of methodological approach 31

3 Qualitative exploration of companies' experience with C2C implementation 35
3.1 Study design 35
    3.1.1 Method and interview structure 36
    3.1.2 Sample selection 37
3.2 Analysing interview results 38
3.3 Deriving propositions for further analysis 47

4 Quantitative exploration of motivational factors and organizational enablers — 51

4.1 Research design — 51
    4.1.1 General research framework and operationalization of variables — 52
    4.1.2 Development of an online survey — 64

4.2 Data preparation for statistical analysis — 69
    4.2.1 Data review and cleansing — 69
    4.2.2 Evaluation of missing Data — 70

4.3 Descriptive analysis — 70
    4.3.1 Company profiles — 71
    4.3.2 Respondent profiles — 77
    4.3.3 Descriptive analysis of main variables — 79

4.4 Exploratory Factor Analysis — 84
    4.4.1 Testing underlying assumptions — 84
    4.4.2 Extraction and factor rotation — 89
    4.4.3 Results of analysis — 90

4.5 Multiple linear regression — 104
    4.5.1 Research framework — 104
    4.5.2 Underlying assumptions for the multiple regression analysis — 105
    4.5.3 Results of multiple regression analysis — 112

5 Discussion of findings — 117

5.1 Determinants of CE adoption — 117
    5.1.1 Driving forces for implementation — 117
    5.1.2 Organizational context and the implementation process — 122

5.2 Building an organizational environment to foster CE innovations — 129

6 Conclusion and implications — 135

6.1 Implications for theory — 135
6.2 Implications for practice — 137
6.3 Limitations and avenues for future research — 141

7 References — 145

8 Appendix — 161

8.1 Appendix A: Online survey — 161
8.2 Appendix B: Descriptive results on responding companies — 172
8.3 Appendix C: Inter-item correlation tables — 174

Index of Figures

Figure 1:	Overview of thesis structure	7
Figure 2:	Typology of eco-innovations	11
Figure 3:	Outline of a Circular Economy	16
Figure 4:	Five-step process for the transition towards eco-effectiveness	22
Figure 5:	Eco-efficiency and eco-effectiveness in the C2C design process	24
Figure 6:	The biological and technical cycle	25
Figure 7:	Exemplary C2C certification product scorecard	28
Figure 8:	Data structure of qualitative results	46
Figure 9:	Conceptual research framework	52
Figure 10:	Conceptual diagram of moderator effect	62
Figure 11:	Structure of the online survey	66
Figure 12:	E-mail invitation for survey participation	67
Figure 13:	Title page of online survey	68
Figure 14:	Overview of survey respondent set	71
Figure 15:	Year in which the company received the first C2C certificate	74
Figure 16:	Future plans to prolong or extend C2C certification	76
Figure 17:	Reasons not to prolong the C2C certification	76
Figure 18:	Age distribution of respondents (years)	77
Figure 19:	Averages for motives for C2C implementation and certification	81
Figure 20:	Review of met or unmet expectations	82
Figure 21:	Final evaluation of C2C experience	83
Figure 22:	Assessment of certification costs	83
Figure 23:	Research framework	105
Figure 24:	Scatterplot to test for homogeneity of residuals	107
Figure 25:	Overview of final results	114
Figure 26:	Simple slope analysis for significant moderation effect	116
Figure 27:	Value Chain framework	131
Figure 28:	The ReSOLVE framework	132
Figure 29:	Share of C2C certified products of respondent companies	172
Figure 30:	Other labels for environmental or social responsibility	172
Figure 31:	Highest certification level of C2C products of respondent companies	173
Figure 32:	Extent of C2C implementation of respondent companies	173
Figure 33:	Inter-item correlations for motivational factors	174
Figure 34:	Inter-item correlations for organizational context factors	175

Index of Tables

Table 1:	Overview of interviews	39
Table 2:	Survey items on satisfaction with C2C implementation	54
Table 3:	Survey items on motivations to implement	56
Table 4:	Survey items on organizational enablers – relationship with certification partner	58
Table 5:	Survey items on organizational enablers – level of implementation	59
Table 6:	Survey items on organizational enablers – technical synergy	60
Table 7:	Survey items on organizational enablers – C2C specific context factors	61
Table 8:	Survey items on organizational enablers – new product success	64
Table 9:	C2C certification level of most of the company's certified products	72
Table 10:	Size and year of foundation of respondent companies	73
Table 11:	Industry and commerce focus of respondent companies	74
Table 12:	Country of origin of respondent companies	75
Table 13:	Tenure of respondents	78
Table 14:	Respondents' functional background	79
Table 15:	Descriptives and correlations for the group of motivational variables	87
Table 16:	Descriptives and correlations for the group of organizational context variables	88
Table 17:	MSA and communalities for motivations – after deletion of M2	91
Table 18:	Factor loadings of motivations – before deletion	92
Table 19:	Factor loadings of motivations – after deletion of loadings < 0.5	93
Table 20:	Final set of motivational factors (ordered by loadings)	94
Table 21:	Results for Cronbach's alpha and CITC for motivational factors	96
Table 22:	MSA and communalities for organizational context variables – after deletion of TS4, CC2 and CC4	97
Table 23:	Factor loadings of organizational context variables – before deletion	98
Table 24:	Factor loadings of organizational context – after deletion of loadings < 0.5	100
Table 25:	Final set of contextual factors (ordered by loadings)	101
Table 26:	Results for Cronbach's alpha and CITC for organizational context factors	102
Table 27:	Factor loadings of satisfaction-related variables	103
Table 28:	Results for Cronbach's alpha and CITC for satisfaction (DV)	103
Table 29:	Reliability of control variable 'new product success'	104
Table 30:	Analysis of multicollinearity	109
Table 31:	Model summary and regression results	113

List of Abbreviations

AVE	Average variance extracted
B2B	Business-to-business
B2C	Business-to-customer
B2G	Business-to-government
C2C	Cradle to Cradle
C2C PII	Cradle to Cradle Products Innovation Institute CE
CEO	Chief Executive Officer
CFA	Confirmatory factor analysis
CITC	Corrected item-to-total correlation
CSR	Corporate social responsibility
DF	Degrees of freedom
DV	Dependent variable
e.g.	exempli gratia (English: for example)
EFA	Exploratory factor analysis
EM	Expectation maximization
EPEA	Environmental Protection Encouragement Agency
et al.	et alii (English: and others)
EU	European Union
F	Test statistic of F-test (F-statistic)
i.e.	id est (English: that is)
IV	Independent variable
KMO	Kaiser-Meyer-Olkin criterion
MBDC	McDonough Braungart Design Chemistry
MCAR	Missing completely at random
MSA	Measure of sampling adequacy
n.s.	Not significant
N/A	Not applicable
PCA	Principal component analysis
R&D	Research and development
R^2	Coefficient of determination (explained variance)
RQ	Research question
SD	Standard deviation
SDG	Sustainable Development Goals
USD	US Dollar
VIF	Variance inflation factor

1 Introduction

> *"Using less of the Earth's resources more efficiently and productively in a circular economy and making the transition from carbon-based fuels to renewable energies are defining features of the emerging economic paradigm in the new era, we each become a node in the nervous system of the biosphere."*
>
> Jeremy Rifkin

It seems an obvious and common understanding that natural resources are a finite good and should be handled prudently. However, our current economic system shows that this simple logic is difficult to realize, which is why the world's resources are constantly being depleted. The awareness of the extensive negative consequences seems to be just dawning for a larger share of consumers, producers or broader society. This finally spurs the transition of sustainability concepts from niche to mainstream, unveiling new chances for our economy. The idea of the Circular Economy (CE), as promoted by the Ellen MacArthur foundation in the last few years, is such a chance.

This dissertation aims to explore the motivations of companies who set out to commit to the Circular Economy path and how they can succeed in this process. The Cradle to Cradle (C2C) paradigm with its certification program, a way of translating the Circular Economy into the implementation level, provides a suitable empirical field to answer the central questions of why and how companies successfully implement the concept. The following introductory chapter illustrates the research relevance (chapter 1.1). Chapter 1.2 elaborates the research aim, broken down into three main research questions and sketches the main contributions. In the final section, the development of the structure of the work illustrates the underlying research design (chapter 1.3).

Introduction

1.1 Research relevance

In recent years, the idea of a Circular Economy has been gaining critical momentum and increased attention. While the concept of sustainability has been discussed and taken up repeatedly in different facets for a long time already, current developments, like the 'Fridays for Future' movement, spur calls for a more radical change of the current economic system, which is characterized by a 'take-make-waste' approach, and lead to a rapidly evolving debate on alternative ways of production and consumption, both in research and practice. A growing number of companies show sustainability adaptations of sourcing and production processes aiming at damage control or subsequent corrections of environmental problems. For instance, the ocean clean-up project[1] was founded with the goal of ridding the oceans of plastics. Also, the variety of companies using recycled materials as input resource is large and seems to further increase. Examples range from established market players like Adidas who produce sports gear and shoes using collected ocean plastic[2] to newer market entrants building their production on the idea of upcycling waste materials, e. g. the Tire Belt Company, a Berlin-based company which uses old tires to manufacture belts[3]. Notwithstanding extant advancements and greater alertness for resources, the concept of Circular Economy pushes the boundaries of many initiatives by presenting an economic model that doesn't produce waste in the first place, combined with the preeminent idea of achieving economic growth nonetheless.

On the business side, the growing customer awareness for regenerative products, healthy materials and sustainable consumption has raised a need for a new imperative in environmental efforts that go beyond the prevailing sustainability activities. This contrasts with an ever increasing global market value of plastics which is reflected by a forecasted growth from 523 billion USD (2018) to over 750 billion USD by 2027 (Statista, 2020a). And despite ongoing efforts regarding waste avoidance, only about one third of the plastic waste collected in the EU was recycled in 2018 (9,4 million metric tons of 29,1 million metric tons) (Statista, 2020b). These exemplary numbers not only underline the urgency for action with respect to a new production and consumption paradigm, but also emphasize the tremendous business opportunities in this field. At this point, the Ellen MacArthur foundation gained an outstanding role in promoting not only the idea of a Circular Economy but the related business imperative for organizations across all sectors (Ellen MacArthur Foundation, 2012, 2016b). The awareness of the Circular Economy paradigm and its relevance for the business world has certainly increased in the last years, e. g. evidenced by strong visi-

---

1 https://theoceancleanup.com/ (accessed 10 October 2020)
2 https://www.adidas.de/en/sustainability-parley-ocean-plastic (accessed 10 October 2020)
3 https://tirebelt.com/ (accessed 10 October 2020)

bility of the issue at the World Economic Forum in Davos (World Economic Forum, 2020). However, the actual realization of a circular economic system appears to still be in its infancy (see chapter 2.2). To further enhance this development, numerous initiatives and programmes were launched. At the European level, for instance, the European commission released the Circular Economy Action Plan in 2015 in order to accelerate the shift towards a Circular Economy (European Commission, 2020), which is supported by an analysis of the Ellen MacArthur foundation that indicates a potential to increase resource productivity in Europe by 3 % annually (Ellen MacArthur Foundation, 2015a). An increased attention and potential pressure from customers is also evident when looking at the number of voluntary sustainability standards, also referred to as ecolabels. Such standards or ecolabels, like the EU ecolabel[4], B-Lab[5] or the environmental management system of the ISO14000 standard[6], have arisen numerously and present a popular means to demonstrate sustainable behaviour for companies (Delmas et al., 2013; Rubik et al., 2008). The positive organizational echo to ecolabels has also been critically discussed since research resulted in equivocal results when looking at the actual increase in transparency and easier comparability of sustainability efforts (Christmann and Taylor, 2006; Simpson et al., 2012).

At the same time, the research discourse on Circular Economy has so far largely focused on definitional and theoretical discussions, leaving the organizational perspective somehow disregarded (Bakker et al., 2010; Neutzling et al., 2018; Schiederig et al., 2012). This could be due to the vague delimitation of Circular Economy to the more general idea of sustainability and sustainable innovation, which has been the subject of controversial research discussions for decades already (Ghisellini et al., 2016; Lieder and Rashid, 2016). Furthermore, a lack of one widely established and assessable definition of the Circular Economy concept (Kirchherr et al., 2017; Blomsma and Brennan, 2017; Geissdoerfer et al., 2016) and a high number of related concepts and notions (Blomsma and Brennan, 2017; Bocken et al., 2017) present another hurdle for research in the organizational context.

In addition to that, the idea of sustainability and many related schools of thought have been regarded as a breach with the established economic system striving for constant economic growth (Bjørn and Hauschild, 2013; Daly, 1997; Ghisellini et al., 2016). This scepticism goes along with a prevailing pursuit of efficiency and focus on improving established practices at the organizational level which might hinder the development of new perspectives on production and consumption patterns (Braungart et al., 2007; Ghisellini et al., 2016). However, by providing new ideas of resource use, ownership and business models, the Circular Economy idea offers a solution

---

4   https://eu-ecolabel.de/en/ (accessed 10 October 2020)
5   https://bcorporation.net/about-b-lab (accessed 10 October 2020)
6   https://www.iso.org/iso-14001-environmental-management.html (accessed 10 October 2020)

to actually combine sustainability-related goals with economic growth (Ellen MacArthur Foundation, 2012; Stahel, 2010; Tukker, 2015). A paradigm shift is hence not only a business opportunity but presents an imperative in the light of depleting natural resources, price volatility and an increasing purchasing power stemming from a growing middle class, especially in emergent markets (Ellen MacArthur Foundation, 2013). Furthermore, such a "recoupling of the relationship between economy and ecology" (Braungart et al., 2007, p. 1338) would allow manufacturers and consumers to shift away from a restriction-driven mindset as production and consumption would not be tantamount to resource exploitation and long-term damage (McDonough and Braungart, 2013). The preeminent idea of eliminating waste by turning it into nutrients pushes the boundaries of prevalent recycling efforts. As such, the Cradle to Cradle (C2C) paradigm presents one approach of operationalizing the CE idea at the organizational level (Bocken et al., 2014; Braungart et al., 2007; Ellen MacArthur Foundation, 2012). By defining concrete product development guidelines on the material level, C2C puts special emphasis on the toxicity of material composites and calls for the use of benign substitutes that allow for a reutilization at the end of the product life cycle without creating harmful waste (Braungart et al., 2007; McDonough et al., 2003).

Another approach to unlock untapped potential in the area of Circular Economy is viewing it through the lens of innovation management. Yet, the discussion on sustainability as a driver for innovation has been around in academia and practice for a long time already. For instance in the work of Porter and Van der Linde (1995), who suggested already in the 1990s that environmental standards can spur innovation. Or Nidumolu et al. (2009) who showed that sustainability initiatives are a key source for "organizational and technological innovations that yield both bottom-line and top-line returns" (Nidumolu et al., 2009, p. 3). However, the actual integration of sustainability into the company's innovation process can be challenging. On the one hand, the innovation process often has to be adapted in a way that sustainability-related factors are included in the very early design phases of the product development process already (Abele et al., 2005; Bocken et al., 2014; Carrillo-Hermosilla et al., 2010). On the other hand, the selected activities might vary depending on the company environment and commitment, ranging from more symbolic acts to profound adaptations of organizational routines, and often require a thorough risk assessment (Hansen et al., 2009; Ketata et al., 2015; Schons and Steinmeier, 2016).

1.2 Research objectives and contribution

Despite the current prominence of the Circular Economy concept in politics, practice and a growing body of research, the organizational acceptance and a substantial uptake of CE-related innovations is not univocally evident (Bocken et al., 2017; Lieder and Rashid, 2016). While research on conceptual facets and definitional subtleties has been a central topic for numerous previous studies (see Ghisellini et al. (2016) for a comprehensive literature review, or Kirchherr et al. (2017) for a meta-analysis of CE definitions), the analysis of organizational processes related to CE implementation has mostly remained on an anecdotal level. Thus, in order to further develop an understanding of a company's CE transition, the organizational processes of initiating and implementing CE-related activities at the organizational level merits further exploration (Blomsma and Brennan, 2017; Lieder and Rashid, 2016; Paramanathan et al., 2004; Schiederig et al., 2012; Tollin and Vej, 2012).

Based on this deficit of research at the implementation level and since intentions don't necessarily always translate into action, so that some companies more than others seem to successfully manage the CE transition, the central aim of the present work is to analyse why and how companies can succeed in the shift towards CE. Taking up the difficulty of a missing clear-cut CE definition, which can be transferred into concrete and comparable organizational activities, the Cradle to Cradle concept serves as a suitable empirical field for this research project. Three main research questions establish the intent of the present work and guide the research project:

RQ1: Why do companies decide to engage in Circular Economy (CE), specifically through the adoption of the Cradle to Cradle (C2C) standard?

RQ2: Which organizational factors are critical for the implementation of C2C and how does the organizational context influence the C2C implementation?

RQ3: How can CE standards like C2C be successfully anchored in a company in the long run?

The approach to answer these questions is twofold and makes use of a hybrid data collection method (Edmondson and McManus, 2007). In a first step, C2C certified companies are interviewed to qualitatively explore their main motives for initiating C2C as well as implementation experiences. Building on the coded results and developed propositions, the second step comprises of a quantitative exploration using an online survey. The data gathered among C2C certified companies builds the basis for an exploratory factor analysis which derives main motivational and organizational factors, and a multiple linear regression to estimate which of these factors influence a company's satisfaction with their C2C implementation.

Employing the hybrid data collection method, this work contributes to the extant research in the field of Circular Economy in three key areas. First, this study advances the CE literature by combining theories from the research fields on sustainability and innovation management in order to quantitatively explore how companies initiate and implement CE innovations using the example of the C2C standard. Especially the primary data collection method in combination with a broad data set has so far been absent in the research field related to CE[7]. Second, since the underlying research model builds on extant theory from related research fields, the results corroborate previous studies on success factors in new product development and complement them with insights particular to circular innovations. Third, a more practical than conceptual perspective is elaborated regarding the organizational context. This would spur the transition to CE by further advancing the state of research from a 'what' focus to answer the 'how' questions.

1.3   Structure of the thesis

In order to realize the research endeavour, the dissertation is organized in six chapters (see Figure 1). After introducing the research relevance and objectives in this first chapter, the theoretical foundation is set by elaborating on the background of sustainable innovations (chapter 2). Hereby, the focus remains on the more recently established concepts of Circular Economy and the selected empirical field of Cradle to Cradle innovations. Given the intermediate state of prior theory, chapter 3 qualitatively explores main motivational forces behind implementation decisions by using the concrete example of C2C and draws a first picture of the companies' experiences during the implementation process. After the literature analysis and the development of a conceptual research framework, chapter 4 introduces the two-step approach to the quantitative analysis before providing the descriptive results of the survey. Reflecting the explorative character of the study, the research design starts with an exploratory factor analysis related to key motivational and organizational factors followed by a multiple regression analysis. The results are consolidated and discussed in chapter 5 in order to derive enabling and hindering conditions in the organizational context. Based on that, the thesis closes by outlining the implications for theory as well as managerial practice. Finally, limitations are listed to point out future avenues for research (chapter 6).

---

7   To the point in time and to the best of the author's knowledge.

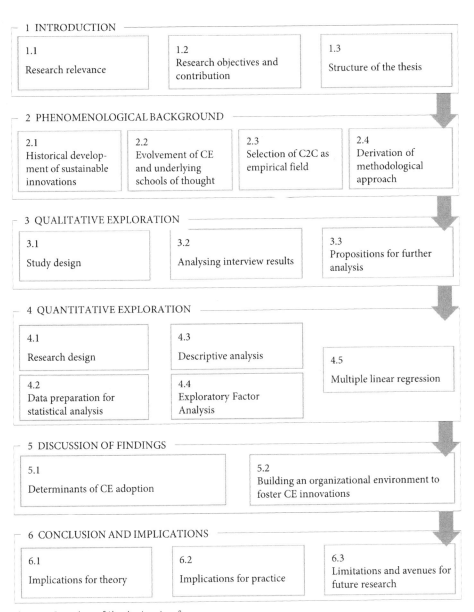

Figure 1: Overview of thesis structure[8]

---

8 Author's illustration.

2 Phenomenological background

While the idea of sustainability and sustainable innovation is not a recent phenomenon in research, the concept of Circular Economy more strongly points at a field of action from a practical perspective and bears additional potential for innovation management.

In the following, the conceptual and theoretical foundation is presented, starting with a short retrospective on the research field of sustainable innovations and evinces a rather fragmented research landscape of related notions (chapter 2.1). Further, the concept of Circular Economy idea is introduced in more detail as it builds the foundation for the underlying work (chapter 2.2). Closely related to CE, the Cradle to Cradle paradigm is presented and its suitability as empirical field elaborated in chapter 2.3.

2.1 Brief perspective on the historical development of sustainable innovations

The concept of sustainability is certainly not a new phenomenon and numerous studies have already researched its origin and historical evolvement. Having its roots in the forestry sector dating back to the 18th century, the idea of sustainability developed in the context of wood harvesting. To avoid wood scarcity, the harvested amount of wood needed to be aligned to the amount that can re-grow (Mantel 1990 in Geissdoerfer 2016). Decades later, the concept sustainable development was the key element of the Report of the World Commission on Environment and Development, known as the Brundtland Commission. The report established the well-known and widely accepted definition of sustainable development as the development: "to ensure that it meets the needs of the present without compromising the ability of future generations to meet their own needs" (United Nations, 1987). This understanding of the concept allows for a variety of implications and can be addressed by many disciplines. To be able to transfer the key points of the definition to the organizational level, researchers have linked the concept of sustainable development with different corporate areas and activities. Just to name a few, there are manifold studies on corporate sustainability (e. g. Van Marrewijk and Werre, 2003), corporate social responsibility (e. g. Carroll, 1999; Carroll and Shabana, 2010), sustainability reporting (e. g. Gray et al., 1995), or corporate citizenship (e. g. Carroll, 1998).

Undoubtedly, the role of innovation is crucial for the realization of sustainable efforts in any manner. However, a clear definition of sustainable innovation is difficult to find and multitudinous definitional subtleties exist[9]. The presence of the numerous terms and the respective nuances of what it includes or not can be explained by the broad interest in sustainability across many research areas, e. g. economic sociology, innovation management or history (Boons and Lüdeke-Freund, 2013; Ghisellini et al., 2016). A large-scale bibliometric analysis of Schiederig et al. (2012) identified different notions that are used by researchers, 'environmental innovation' being the most frequently used (41%), followed by the terms 'sustainable innovation' (21%) and 'eco-innovation' (18%). Regarding the meaning and implications, the notions don't differ significantly and are often used interchangeably. Of these most established notions, the concept 'eco-innovation' is considered to be one of the most elaborated ones (Schiederig et al., 2012). On the one hand it contains the term 'innovation', which implies a new technology or market (Abernathy and Clark, 1985). On the other hand, it refers to the reduction of a negative environmental impact, due to environmental or other reasons (Carrillo-Hermosilla et al., 2010).

Hitherto, the research field of sustainable innovation is characterized by many dichotomous discussions about whether sustainability targets can be implemented or not, by this discounting a gradual perspective on processes and stepwise approaches towards sustainable innovations (Neutzling et al., 2018).

In their comprehensive study on eco-innovations, Carrillo-Hermosilla et al. (2010) illustrate the variety of implications with the help of a scale, underlining the possibilities for companies to engage in sustainability through eco-innovations and the corporate areas being affected (see Figure 2). The scale classifies types of eco-innovation depending on their radicalness of the produced technological change (incremental vs. radical) and the levels of impact to the system. Moreover the perspectives of the economic as well as the environmental and social sustainability of the

---

9   In 1995, Joseph Huber introduced three key strategies related to sustainable development, which are a substantial building block in the discussion of different perspectives on sustainable development: the sufficiency strategy, efficiency strategy and consistency strategy. While the first and second rather stand for self-limitation and increased efficiency, the latter is strongly interlinked with the circularity idea (Huber, 1995).

system are included. The eco-innovation type tackling the minimization of negative impacts on the system is often based on an incremental system change and is described as *component addition*. Although the addition of components can lead to some improvements, e. g. fewer emissions or better air and water quality, such technologies are not sufficient to overhaul the existing systems. Still, such innovations can help to gain time for the process of developing more radical solutions (Carrillo-Hermosilla et al., 2010). Reflecting eco-innovations following the eco-efficiency idea, the sub-system change can be considered an intermediate step on the way towards the system change. Goods are produced by using fewer resources, resulting in more sustainable products, e. g. through the reduction of negative externalities (Carrillo-Hermosilla et al., 2010; Kobayashi et al., 2006).

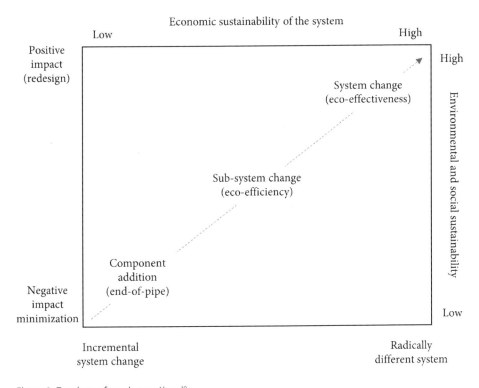

Figure 2: Typology of eco-innovations[10]

---

10   Author's illustration based on Carrillo-Hermosilla et al. (2010).

Having a high impact on economic, environmental and social sustainability through a radically different system, eco-innovations can also lead to an entire *system change*, which implies the creation of positive effects (Carrillo-Hermosilla et al., 2010). This eco-effective solution is linked to the principles of Industrial Ecology, where production of goods is incorporated in closed-loop systems and waste can be further used as a resource for new products (Carrillo-Hermosilla et al., 2010). Based on these characteristics, the system change innovations are strongly linked to the concept of Circular Economy. Even though the typology offers a holistic overview of the differences in eco-innovative approaches, there is no uniform solution for a component addition, sub-system or system change. As organizations are subject to specific circumstances of their industrial system, e. g. the economic development of a country or differences in customer perceptions, the presented typology does not present a homogenous perspective on eco-innovations. Carrillo-Hermosilla et al. (2010) conclude that eco-innovations still depend on spatial, temporal and cultural characteristics of an organization.

How organizations can profitably integrate sustainability into their innovation processes was researched by Hansen et al. (2009) with the help of a Sustainability Innovation Cube, a framework that considers the business opportunities created through sustainability-oriented innovations (SOI)[11] and the risk involved in such efforts. Even though the integration of sustainability into innovation efforts can provide new business opportunities, such as new customer segments or a new source of ideas, decision-makers are confronted with risks that are likewise connected to such an integration. In particular, there are potential negative impacts on the environment as well as the even more insecure customer perception which are difficult to foresee for decision-makers due to the newness of ideas or technologies (Hansen et al., 2009). The concept of the sustainability innovation cube underlines the special complexity which goes along with sustainable innovations, particularly in the early phases of the innovation process when special knowledge and a holistic perspective on environment, economy and society is required (Bocken et al., 2014; Hansen et al., 2009; Ketata et al., 2015). Despite the high complexity, researchers have also argued that there is a significant business opportunity behind the integration of sustainability aspects into an organization's innovation process (e. g. Schaltegger and Synnestvedt, 2002; Senge, 2008).

The integration of sustainability aspects in the innovation process also implies additional challenges that companies and decision-makers have to face in the light

---

[11] In a more recent publication, Adams et al. (2018) provide a detailed literature review of over 100 articles on SOI which, amongst other findings, confirms the early stage of its theoretical development and further points at the necessity to develop practical tools and methods in order to accelerate practical acceptance and implementation.

of growing demand for companies to cope with societal and environmental challenges. The body of research with respect to the challenging trade-offs businesses have to cope with when integrating mixed motives has been accurately synthesized by Margolis and Walsh (2003), who point at the duelling demands with respect to financial versus societal expectations. This controversy also poses a challenge for the research agenda. The related research streams are also manifold, ranging from supply chain management to marketing or business model innovation, and emphasize the transdisciplinary character of sustainable innovations (Tollin and Vej, 2012).

2.2 Evolvement of the Circular Economy concept and underlying schools of thought

The idea of a Circular Economy as a contrast to a linear economic system is not a new idea as such. For decades, scholars have discussed aspects and interconnections and notably potential implications on environment and economy. With the growing academic interest in the topic, particularly in the last years a number of profound and accurate reviews have been developed to reflect the academic discussion on Circular Economy in the business context. For instance, in their extensive literature review on Circular Economy, Ghisellini et al. (2016) identify various theoretical foundations such as Environmental Economics or Industrial Ecology.[12] Hence, it proves difficult to identify a clear origin of the Circular Economy idea. However, the underlying common ground is the explicit counter-position of this idea to the prevalent economic system dominated by a linear flow of materials, described as the 'take-make-waste' paradigm (Bocken et al., 2017; Ghisellini et al., 2016). Different scholars emphasized various elements building the Circular Economy. For example the idea of a shift away from selling products to providing services was profoundly analysed by Stahel, already starting in the 1980s and 1990s (Stahel, 1982). Building on the idea of the combination of tangible products and intangible services, Tukker (2004) further elaborated the concept by developing different types of Product-Service Systems (PSS). A comprehensive review of PSS in the context of Circular Economy (Tukker, 2015) ascertained the growing interest of the research and business community and outlined different design methodologies and their importance in overcoming implementation barriers of PSS for businesses. Further related research streams are, e.g. Industrial Ecology (Ehrenfeld, 1997), Biomimicry (Benyus, 1998), eco-innovations (Carrillo-Hermosilla et al., 2010), the method of Life-Cycle assessment (see LeGwin

---

12 Lieder and Rashid (2016) have comprehensively reviewed the state of CE research as well as its implications for the manufacturing industry. Furthermore, a detailed review of more than 100 CE definitions has been recently developed and published by Kirchherr et al. (2017).

Phenomenological background

(2000) for an elaboration on the theoretical foundation) or the Cradle to Cradle paradigm (McDonough and Braungart, 2002b). In their extensive review of the academic discussion on CE, Ghisellini et al. (2016) conclude that CE can be "a way to design an economic pattern aimed at increased efficiency of production (and consumption), by means of appropriate use, reuse and exchange of resources, and do more with less." (Ghisellini et al., 2016, p. 18). The variety of related definitions has certain overlaps as well as delimitations. Accounting for this multiplicity of related concepts, Circular Economy can be regarded as a conceptual umbrella term, hence it can emphasize shared characteristics without compromising already existing phenomena (Blomsma and Brennan, 2017; Hirsch and Levin, 1999). In their analysis, Blomsma and Brennan (2017) further emphasize the stage of the CE research landscape.

Given the current state of the theoretical and practical debates on CE, the classification as an umbrella term can lead to a "catalytic function", hence leverage the development and diffusion of the idea (Blomsma and Brennan, 2017). Furthermore, the authors present a typical development pathway of umbrella concepts based on the work of Hirsch and Levin (1999) and position the CE concept according to its development stage. Starting in the 1960ies in the preamble stage, the Circular Economy concept further passed the excitement period (1985-2013) and is presently classified in the so-called *Validity Challenge* Period. This stage is characterized by a controversial discourse on conceptual or paradigmatic ambiguities and leads to the necessity for further elaboration of the concept, in the *Further Work* period. Only then, it can be determined whether the concept is robust enough and evolves into the *Coherence* stage, is classified as *Permanent Issue* or even falls away in the *Construct Collapse* stage (Blomsma and Brennan, 2017; Hirsch and Levin, 1999).

In the identified stage of challenging the validity of the CE concept, the approach of the Ellen MacArthur Foundation has gained considerable prominence during the last years. Founded in 2010, the British professional long-distance sailor and holder of the world record for the fastest solo circumnavigation of the globe, Ellen MacArthur, founded the Ellen MacArthur foundation "to accelerate the transition to a circular economy" by "establishing the circular economy on the agenda of decision-makers across business, government, and academia." (Ellen MacArthur Foundation, 2016a). Looking at the development of CE in the academic field, the interest has substantially increased, especially during the last decade. Compared to the year 2010 with 10 scientific publications, the number has almost tripled in 2015 with 27 publications (Lieder and Rashid, 2016). In their bibliometric analysis, Lieder and Rashid (2016) also identify a strong geographical focus on China which results from passing the Circular Economy Promotion Law and a national CE indicator system.

2.2.1 Main principles and core ideas

As described earlier, the variety of definitions for Circular Economy related concepts and their respective emphasis on different cornerstones is multifold (see Blomsma and Brennan (2017), p. 605 for a graphical review of different CE interpretations). One of the most established definitions, which has also been adopted by the World Economic Forum, has been published by the Ellen MacArthur Foundation (2012). They define Circular Economy as "an industrial system that is restorative or regenerative by intention and design. [...] It replaces the 'end-of-life' concept with restoration, shifts towards the use of renewable energy, eliminates the use of toxic chemicals, which impair reuse, and aims for the elimination of waste through the superior design of materials, products, systems, and, within this, business models" (Ellen MacArthur Foundation, 2012, p. 7).

The idea of eliminating the concept of waste by establishing the concept of restoration can be realized through the introduction of two value cycles, in which biological and technological materials are intended to flow. The cycles consist of several inner circles which are an important element to ensure value creation as it is assumed that these inner circles lead to higher cost and material savings. It is critical to keep the cost for collection, reprocessing and return of a product into the economy lower than the linear alternative in order to leverage the economic benefits of the Circular Economy (Ellen MacArthur Foundation, 2012). In addition, the idea presented in the so-called CE system diagram suggests keeping materials and components of products as long as possible within the circles, this applies particularly to the technical cycle. However, the positive effect of prolonging product lifecycles and enabling companies to make less use of virgin input materials can be nullified if the longer life cycles prevent companies from using new technologies or innovations which might increase efficiency or allow for new and better materials and processes (Ellen MacArthur Foundation, 2012). To capitalize on the abovementioned positive effects, companies also need to consider the accurate use of materials and components across different product categories as well as the design of products. The cascaded use of materials is an important lever if the full recovery of an original material is too complex or demands too much energy or cost. In this case, materials should be used as input for other product categories in a way that the marginal costs are kept as low as possible. This also addresses the product design which is critical with regard to the use of pure materials and a modular product type. Thus, the decomposition and re-utilization can be more effective and generate higher value for the organization (Ellen MacArthur Foundation, 2012). Following these conditions, the idea of a Circular Economy presents a means to decouple production and consumption from the prospect of economic growth (Ghisellini et al., 2016).

Phenomenological background

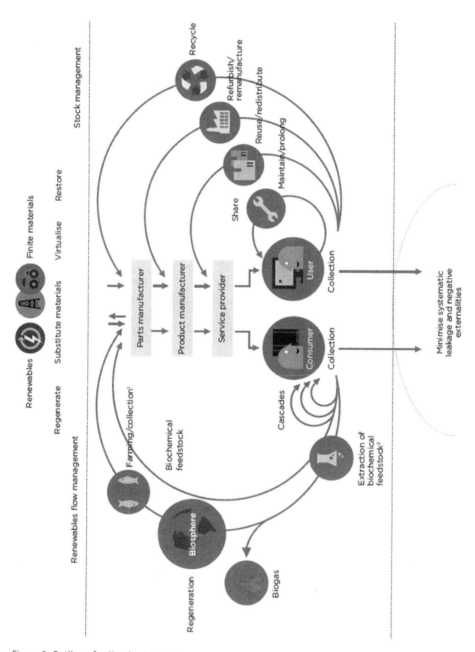

Figure 3: Outline of a Circular Economy[13]

---

13   Source: Ellen MacArthur Foundation (2015a).

The described diagram (see Figure 3) was introduced in the early works of the Ellen MacArthur foundation to illustrate the idea behind the biological and technical cycle and has been developed further, e.g. by adding three underlying principles reflecting three stages of the cyclical approach. The first one calls for the preservation of finite and natural capital which serves as input into the cycles. The second principle addresses the actual cycling by requiring enhanced circulation processes, ensuring re-utilization of materials by keeping them on a constantly high quality level. Finally, the third principle postulates a system effectiveness by eschewing negative externalities (Ellen MacArthur Foundation, 2015a).

In order for the Circular Economy to succeed, the involved actors need to be considered as well. This could imply, for instance, alternative employment models, adapted business models focusing on product-service systems and a new generation of product designers and service providers as well as ambassadors of CE in politics and decision-making (Ellen MacArthur Foundation, 2012; Ghisellini et al., 2016). Especially at the organizational level, this holistic system thinking approach requires the interplay of many different actors and stakeholders. The successful transition towards a Circular Economy also hinges on the emergence of a new generation of innovative designers as the product design plays an essential role. The recent research of Den Hollander et al. (2017) on the crucial role of product design for the CE underlines the complexity of the current transitioning phase. The authors review various tools, methods and design strategies and conclude with the necessity of redefining the existing perspective on product lifetime and end of life by introducing the concept of 'product integrity'. The concept is defined as "the extent to which a product remains identical to its original (e.g., as manufactured) state, over time" (Den Hollander et al., 2017, p. 519) and a typology provides different design approaches to work towards the CE objective of waste elimination.

2.2.2 Critical discussion of the CE concept

As stated above, there is scepticism about the newness or actuality of the Circular Economy idea across several academic discussions. Furthermore, critics question the necessity of compiling different elements of already extant concepts into something new as this can lead to oversimplification. While specific disciplines analyse different CE elements in a more detailed way, e.g. from a technical or chemical perspective, contemplating the waste management or the remanufacturing process, the holistic approach, as enforced by the Ellen MacArthur Foundation, might appear too schematic (Zink and Geyer, 2017). This argument underlines the observation that researchers from the academic field, so far, have remained on a rather superficial discussion level and did not succeed in generating significant insights on concrete implementation activities at the organizational level (Lieder and Rashid, 2016). Furthermore, it is criticized that many analyses neglect the economic concerns which are critical for the numerous industrial actors required for the shift towards a CE. As industrial players base their decision-making on business considerations, this lack of research efforts addressing economic benefits or value creation through CE appears to be a hindering factor for broader acceptance of the concept (Lieder and Rashid, 2016). Another concern raised by Zink and Geyer (2017) questions the actual realization of the CE ideas. According to their argumentation, the circular illustration neglects the existence of markets for primary and secondary goods and materials. Thus, secondary materials which the CE concept aims to establish in order to reduce demand for virgin materials, compete with primary materials in the respective markets, e.g. end-of-life goods or scrap. It is unclear in how far this complexity is considered in the schematic elaboration of CE ideas. Zink and Geyer (2017) also argue that the differentiation of primary and secondary markets raises the question of a potential rebound effect which might over-compensate the positive effects, created by the integration of eco-effective considerations into manufacturing. Their argument builds on the possibility of creating new markets which did not exist before, such as for refurbished smartphones. In this case, customers who cannot afford primary products such as smartphones get access to secondary products, which creates a new market where previously there was none. Thus, the refurbished phones present an addition, not a substitution, creating a rebound effect (Zink and Geyer, 2017).

2.2.3 The CE as an imperative for future business

In parallel to the critical discussions, proponents of the Circular Economy idea advocate the urgency for action. This argumentation is driven by researchers as well as industrial players, public institutions and policy-makers, with the Ellen MacArthur Foundation playing a pioneering role. In its early reports, the foundation underlined the necessity of a shift towards CE with analyses on the value creation throughout different sectors. Based on the developments of worldwide population growth and the increasing number of people having access to more goods and services due to a growing middle class, the current economic system is subject to new risks and challenges as well as an expected increase in the demand for manufactured, packaged goods, replacing the formerly consumed loose or unbranded ones (Ellen MacArthur Foundation, 2013). The resulting demand increase for finite materials comes along with high price volatility and increased pressure on industrial players. Furthermore, the rising prominence of the environmental concerns on the agenda of policy-makers and governments such as the European Union, amplifies the need for a new economic system that promotes the idea of economic growth and sustainability aspects not being mutually exclusive (Ellen MacArthur Foundation, 2012; Ghisellini et al., 2016; Lieder and Rashid, 2016). In the analysis of the current manufacturing and recovery processes, the Ellen MacArthur foundation states that in Europe currently only about 40 % of the generated waste of 2.7 billion tonnes can be reused, recycled or composted, one important issue also being that only few material types allow for an adequate waste recycling. This leads to a recovery rate that is much lower than primary manufacturing rates (Ellen MacArthur Foundation, 2012). In the light of increased resource prices and price volatility, the utilization of waste as a resource can present an important source for competitive advantage for a single company or a whole industrial economy (Ellen MacArthur Foundation, 2013; Ghisellini et al., 2016). Especially for manufacturing companies, the competitive situation with regard to finite resources has gained importance together with the necessity to comply with constantly increasing obligations concerning social and environmental regulations (Lieder and Rashid, 2016). In an extensive analysis of the consumer goods sector, the Ellen MacArthur Foundation has identified that in 2011 about 80 % of the fast moving consumer goods (FMCG) were not recovered, resulting from losses across various steps of the value chain, starting with a value loss in the agricultural supply such as spillage during harvest, to value losses during production, distribution and recovery (Ellen MacArthur Foundation, 2013). In this context, the report also emphasizes the role of the consumer during the use phase, demonstrated by the high rate of around 50 % of food which an average family in the US throws away (Ellen MacArthur Foundation, 2013). This observation underlines the upcoming challenges regarding lost value in the light of the constantly growing world population and an accelerated surge of middle-class growth.

Policymakers and industrial players recognized these imperatives, especially in the recent decade and positioned the CE topic on the agenda of powerful decision-makers. Two prominent examples are the Project MainStream, initiated by the World Economic Forum, which addresses the top executives of companies and aims at establishing CE innovations on a large scale (World Economic Forum, 2020) and the Sustainable Development Goals (SDG), elaborated by the United Nations (UN), out of which several address the issues that are also tackled by the CE concept, hence promoting a global development aligned with CE standards,. For instance, SDG 12 'Responsible consumption and production' or SDG 14 'Life below water' advocates the protection of oceans as well as marine and coastal ecosystems, e. g. through the reduction of plastic packaging[14].

---

14 https://www.un.org/sustainabledevelopment/sustainable-development-goals/ (accessed 08.11.2019)

However, one critical factor impeding the broad implementation of CE approaches, is the system-level perspective. Building on the idea of a holistic system change, the CE idea requires the collaboration of various partners, who have not necessarily worked together yet (Seebode et al., 2012). The required integration of numerous stakeholders, including all players related to the value chain, but also competitors as well as customers, presents the main hurdle when it comes to concrete CE activities, especially on a larger scale. However, it is precisely this systematic approach that could finally induce the long-discussed change as the integrated perspective has been largely missing in previous attempts to bring together social and environmental aspects with business considerations (Lieder and Rashid, 2016).

2.3 Selection of C2C as empirical field

The academic discussions and reviews of the Circular Economy concept underlined the existence of multifold definitions and understandings, not only in academia but also in practice. While this can be regarded as a positive development with respect to the diffusion of the concept, it can also impede the research efforts and at the same time dilute relevant CE activities due to missing comparability, especially concerning company efforts. To address this constraint in the following research project, the selection of the empirical field aimed at providing a more precise and common ground for the understanding of focussed CE efforts. From the related schools of thoughts, the idea of C2C with its corresponding certification scheme provided a solid basis for analysis for numerous reasons which will be detailed in the following.

2.3.1 Origin and development of C2C

The Cradle to Cradle (C2C) paradigm can be regarded as an underlying school of thought of the Circular Economy. Essentially coined and made prominent through the work of US American architect William McDonough and German chemist Michael Braungart, C2C strongly builds on the idea of two cycles and the distinction of materials in order to maintain the quality of a product and its components. The idea of C2C goes back to the early 1990s, when Braungart and McDonough published the "Hannover Principles" as a guideline for sustainable design principles in preparation for the world fair Expo 2000 (McDonough and Braungart, 2012). In 2002, the book 'Cradle to Cradle: Remaking the Way We Make Things' was successfully published and translated into several languages[15]. "Cradle-to-cradle design enables the creation of wholly beneficial industrial systems driven by the synergistic pursuit of positive economic, environmental and social goals. [...] Cradle-to-cradle design defines a framework for designing products and industrial processes that turn materials into nutrients by enabling their perpetual flow within one of two distinct metabolisms: the biological metabolism and the technical metabolism." (Braungart et al., 2007, p. 1343).

By contrasting upcycling and downcycling, hence breaking with the dominant term of recycling, the C2C concept presents a new way of looking at the quality of materials and components through their lifecycles (Braungart et al., 2007)[16]. This is presented as a necessary 'change of perspective' in several ways. On the one hand, the phrase 'cradle to cradle' focuses on the imperative of circularity and emphasizes the counter-positioning to the prevalent linear economic system described as 'cradle to grave' which stands for the prevalent economic system of *making-using-disposing*. On the other hand Braungart and McDonough argue that the widely established idea of efficiency corresponds to 'doing things the right way' whilst C2C underscores effectiveness being more worth pursuing as this would mean 'doing the right things' (Braungart et al., 2007). Finally, the concept further elaborates the idea of the triple bottom line by introducing the triple top line (McDonough and Braungart, 2002a). Linking sustainability efforts with the business context, the triple bottom line framework shall help decision-makers with the evaluation of social and environmental considerations in the pursuit of economic goals (Elkington, 1994, 1997). Braungart and McDonough position the concept of the triple top line as a different perspective that businesses need to take in order to recognize the major business opportunities that are opened up when social and environmental aspects are regarded as potential value leading to greater synergies (Braungart et al., 2007; Lieder and Rashid, 2016). The focus of the triple top line is the front-loading of strategic decisions in order to move "accountability to the beginning of the design process, assigning value to a multiplicity of economic, ecological and social questions that enhance product value" (McDonough and Braungart, 2002a, p. 252).

A clear focus of C2C is to look at products from a chemical perspective in order to identify ideal material compositions based on toxicological characteristics for the integration into the cycles (Braungart et al., 2007). The precision of components and

---

15 In 2019, the book was re-released as part of the Penguin Patterns of Life series, a series of special editions of the best popular science books to explore the patterns of the planet (https://www.penguin.com.au/books/cradle-to-cradle-9781784873653).

16 A first descriptive review of C2C publications and preliminary findings on C2C implementation was elaborated in a previously published working paper by the author of this thesis (Geng and Herstatt, 2014).

materials on the product-design level is suggested to follow a five-step approach (see Figure 4). The first step '*Free of...*' is about the exclusion of materials with heavily negative effects on the environment or human health. This activity can be challenging as many companies have a complex supply chain and do not have the information about the precise compositions or substances. Another hurdle in this first step can be to find an adequate replacement for the toxic substances, which can ensure that the product characteristics are maintained.

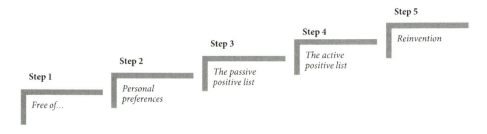

Figure 4: Five-step process for the transition towards eco-effectiveness[17]

After the removal, the next step '*Personal preferences*' suggests to actively define of which materials the product should consist. This implies a profound knowledge about "a substance's toxicological profile and its fate throughout the life cycle" (Braungart et al., 2007, p. 1344). Based on the difficulty of a substantial analysis including all future effects, the step calls for a reasonable judgement and consideration of different trade-offs when selecting the right materials. The third step '*The passive positive list*' supports such trade-off decisions and suggests to systematically classify product materials in order to identify the ones that are best suited to be fed into the two cycles. In this step, the term passive refers to the idea of substances that are identified as critical, however can't be easily replaced. Thus, the third step advocates a profound knowledge about materials and their effects as well as an accurate handling of the critical ones. '*The active positive list*' builds on that by endorsing the "optimization of the passive positive list until each ingredient […] is positively defined as biological or technical nutrient" (Braungart et al., 2007, p. 1344). The activities of this fourth step aim at the optimization of all substances contained in a product after the optimization potential has been identified in the previous step. Finally, once the optimization has been realized on the materials level, the last step '*Reinvention*' addresses the systematic perspective of the C2C idea which considers industrial systems. This steps needs close collaboration with the customer in order to be able to identify for which purposes the products are actually used and how they are disposed of. This

---

17   Author's illustration based on Braungart et al. (2007).

2.3.2 Main characteristics of C2C

The introduction of eco-effectiveness

The C2C paradigm introduces the idea of eco-effectiveness, in contrast to the more established idea of eco-efficiency (McDonough and Braungart, 2002b). Based on the work of Schaltegger and Sturm (1989), the term eco-efficiency was mainly established in the 1990s (Reijnders, 1998; Schmidheiny, 1992) and has been a widely published topic in the area of sustainability until today (Caiado et al., 2017). The most prominent definition was given by the World Business Council for Sustainable Development (WBCSD) in the context of the Rio Earth Summit 1992 stating that eco-efficiency can be achieved by "the delivery of competitively priced goods and services that satisfy human needs and bring quality of life, while progressively reducing ecological impacts and resource intensity throughout the life-cycle to a level at least in line with the Earth's estimated carrying capacity" (Schmidheiny, 1992). Juxtaposing this endeavour, the concept of eco-effectiveness turns away from the target of reducing negative externalities. Instead, there should be a prospect positive contribution to ecological systems and the reduction of waste is superseded by the idea of fully eliminating it through a retained resource productivity and quality (Braungart et al., 2007). The eco-effectiveness approach goes beyond the objective of optimizing existing methods, hence calls for the development of new systems of material flows and the re-design of current processes as well as consumption patterns (McDonough and Braungart, 2002a, 2002b). In this context, the example of the automotive sector applies, when most of the effort has been put into the reduction of negative externalities, i.e. decreasing emissions and higher fuel efficiency. However a radical new approach towards mobility could actually generate positive effects and present a new business opportunity for automotive players (Boons and Lüdeke-Freund, 2013; McDonough and Braungart, 2002b). Such a constant effort to optimize extant processes could eventually lead to a "false sense of security" (Bjørn and Hauschild, 2013, p. 324) and positive implications could be over-compensated through increased use or consumption as described by the rebound effect (Berkhout, 2011; Carrillo-Hermosilla et al., 2010; Greening et al., 2000).

According to the C2C standard, product components circulate in the predefined material flows in a constant productivity level, thus enabling "materials to maintain their status as resources and accumulate intelligence over time (upcycling)" (Braungart et al., 2007, p. 1338). While eco-effectiveness, in contrast to eco-efficiency, doesn't aim for improvements of the current system and suggests a new approach towards the design of product life cycles, the two strategies should not be regarded as mutually exclusive but rather as complementary (see Figure 5), particularly in a transitioning phase.

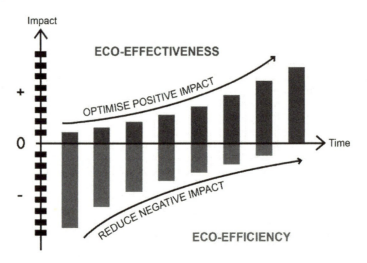

Figure 5: Eco-efficiency and eco-effectiveness in the C2C design process[18]

Such a transition shall be guided by three core principles of the C2C idea. The first *Waste equals food* summarizes the key target of used products, materials and resources such as water, which are traditionally treated as waste and are being disposed of as new resources which can be so-called nutrients for new products, both in the technical and the biological cycle. Hereby the recovery of rare materials is particularly critical. Secondly *Use current solar income* specifies energy use to be location-specific in order to leverage natural and regional energy flows, including solar, wind and water. This principle was re-named into *Use Renewable Energy* to clarify the inclusion of all renewable energy sources as alternative to fossil energy use (MBDC LLC, 2012). The third pillar *Celebrate diversity* comprises creative and customized solutions instead of one-size-fits-all solutions, e.g. renting of washing machines with a customized detergent that fits the local water characteristics. It also addresses the importance of technology diversity in order to foster innovations and creative solutions to realize C2C cycles (MBDC LLC, 2012; McDonough et al., 2003). Through the accurate analysis of a product environment, the company takes into account the local social, cultural and economic factors and thus can build a form of resilience (McDonough and Braungart, 2002b).

---

18   Source: EPEA (http://braungart.epea-hamburg.org/en/content/c2c-design-concept, accessed 13.04.2018).

Biological and technical cycles

The C2C concept divides product materials into biological and technical nutrients which are fed into two metabolisms, namely the biological and technical cycle (Figure 6). According to their ability to be integrated in one of the cycles after their end of life, all materials and product components need accurate selection prior to production (Braungart et al., 2007).

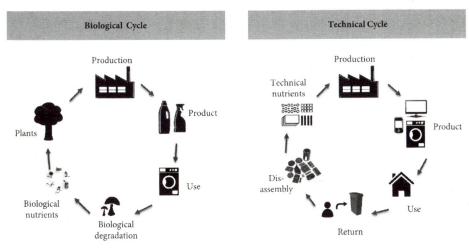

Figure 6: The biological and technical cycle[19]

The biological metabolism involves biodegradable, i. e. compostable materials, which can be natural or plant-based and are sold to the customer as *nondurable goods of consumption* (Braungart et al., 2007; Braungart and McDonough, 2011). The main characteristic of such materials is that they don't pose any risk for the environment, neither during the use phase nor at their end-of-life stage, and that they need to be returnable to natural systems. Thus, also synthetic materials like biopolymers are classified as nutrients for the biological cycle. Examples are products that are biologically, chemically or physically changed by the customer, such as textiles for clothes or furniture and brake pads for cars or bikes (Braungart et al., 2007).

The technical metabolism applies to *durable goods of service* that need disassembly after their return from the customer and are not biodegradable. The aim is to decompose the product components in such a way that all materials can be endlessly reused in new products without losing quality and ultimately even improving their performance through their constant appliance (Braungart et al., 2007). Typical examples of products circulating in the technical metabolism are home appliances

---

19 Author's illustration based on Braungart et al. (2007).

such as washing machines as well as other electronic devices. Many product categories can also be attributed to both cycles as they comprise technical and biological components, e.g. office chairs contain degradable textiles that need to be fed into the biological cycle as well as disassemblable plastic or metal parts that return into the technical cycle. Particularly the idea of the technical cycle poses high demands on product development and particularly the chemical and technical requirements of materials and resources (Lieder and Rashid, 2016). It also challenges existing consumption and ownership patterns as well as business models as the manufacturers retain control and possession of the products in order to be able to leverage the technical cycle (Bjørn and Hauschild, 2013; Braungart et al., 2007). In a more advanced scenario, components and materials don't have to be returned to the manufacturing company but can be managed and distributed through an 'intelligent materials pooling' system (Braungart et al., 2007). Such a network of involved value chain actors would be responsible for the information and knowledge management on materials and could distribute needed components according to pre-defined criteria so that a proper recyclability in the sense of upcycling can be ensured and reuse can be optimized (Braungart et al., 2007). However, the idea of such a material bank has remained a theoretical concept so far.

2.3.3 C2C on the organizational level: the certification program

The idea of transferring the C2C standard into the business context goes back many years. In the late 1980s, Michael Braungart founded the 'Environmental Protection Encouragement Agency' (EPEA) in Hamburg. Few years later, McDonough and Braungart established the 'McDonough Braungart Design Chemistry' (MBDC). Both organizations aimed at the diffusion of the C2C standard in the business environment, which in the year 2005 resulted in the creation of the 'Cradle to Cradle Certified' program offering product-specific certifications (Bakker et al., 2010; Cradle to Cradle Products Innovation Institute, 2014; Toxopeus et al., 2015). The certification scheme covers all industries and geographical areas.

The standard builds on five categories that are assessed during the certification process[20]. The category *Material Health* focuses on the constituent components of

---

20 At the end of the present research project, a new version of the C2C certification scheme, the "Cradle to Cradle Certified Version 4.0", was developed and the certification categories adapted. The categories 'material health' and 'social fairness' remained unchanged. The other categories were amended and renamed into 'Product Circularity', 'Clean Air & Climate Protection' and 'Water & Soil Stewardship'. Further, the certification level 'basic' is not part of the certification scorecard any more. (https://www.c2ccertified.org/get-certified/cradle-to-cradle-certified-version-4, accessed 13 March 2021).

a product. Companies get a list of materials which have been chemically based on their potential negative impacts on human health and environment. The aim is to fully avoid the materials classified as harmful in the so called of chemicals and replace them with alternatives that comply with the ard. The second category *Material Reutilization* addresses the product's potential to become a nutrient in the respective cycle once the product lifecycle has ended as well as the manufacturer's recovery efforts. In the third category *Renewable Energy and Carbon Management* the amount and nature of energy used for production are evaluated, promoting the prevalence of renewable energy sources. Another resource focus lies in the category *Water Stewardship*, in which companies are assessed based on the water consumption and affluent resulting from the product manufacturing. The last category *Social Fairness* addresses the work conditions related to the production and measures how companies meet the predefined requirements for production facilities and general environmental impacts (MBDC LLC, 2012). A certification level is granted depending on the evaluations in the presented categories, whereby each product must be evaluated separately, since there is no certification scheme for processes or whole companies so far. The progressive certification logic begins with the Basic level, which is considered a provisional step on the way towards C2C-conform products. Then the levels Bronze, Silver, Gold and Platinum follow, providing companies with a progressive certification logic. As the certification is granted on a product-level, one company can hold different levels for different products. Irrespective of the level, every certification must be renewed every two years, proving a certain improvement in the C2C certification criteria (Cradle to Cradle Products Innovation Institute, 2014).

The C2C Products Innovation Institute (C2C PII), the organization that administers the C2C product standard and assists companies with the C2C certification, currently lists over 500 certified products on their website (Cradle to Cradle Products Innovation Institute, 2019a). The certification is available across various industries and applying to B2B as well as B2C sectors (Cradle to Cradle Products Innovation Institute, 2014). The industry with the highest share of product certifications is summarized under the classification 'Interior Design & Furniture' and makes up around 40% of the total amount of certified products. It covers 210 certified products, such as carpeting, other floor coverings, ceiling products, and office furniture. The second highest number of certifications lies in the sector 'Buildings Supply and Materials', containing over 170 products, e.g. building exteriors, sunscreen or roller-shade systems, glass, and insulation materials. These categories are followed by the category 'Home & Office Supply' (67 certifications) and 'Fashion & Textile' (42 certifications). Particularly in the last years, the latter has gained increasing attention and was the subject of numerous projects such as the 'fashion positive' initiative[21]. The growing prominence was accompanied by certifications of well-known textile brands in the business-to-customer field, ranging from more affordable apparel lines by Lidl or C&A to premium labels like Wolford[22].

Examples from the aforementioned products groups presenting a large share of C2C certifications are numerous and range from niche to mainstream. The C2C PII published a report called "Innovation stories" in 2013, containing 60 examples of certified companies across many industries and geographical locations and thus providing insights into the potential sectors and product ranges of C2C certification (Cradle to Cradle Products Innovation Institute, 2013).

In order to get a certification, companies must prove their compliance with the certification requirements in the respective categories (see Figure 7).

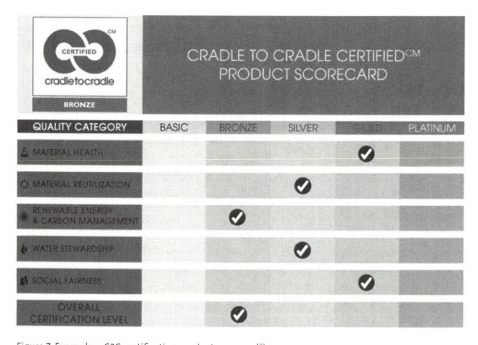

Figure 7: Exemplary C2C certification product scorecard[23]

A number of accredited assessment bodies are eligible to grant the certification. Companies can currently select between ten assessors in Europe, three in North America and two in South America (Cradle to Cradle Products Innovation Institute,

---

21   https://fashionpositive.org/
22   See the product registry of the C2C PII for the complete list of certified companies: https://www.c2ccertified.org/products/registry
23   Source: https://www.c2ccertified.org/get-certified/levels (accessed 03.04.2019).

2019b). The C2C PII hereby functions as an administrator and provides the guidelines of the *Cradle to Cradle Certified™ Product Standard*. In addition, the institute is responsible for the actual certification process, once companies have completed the required reports. In order to analyse and adapt a product according to the C2C standard, companies often reach out to the assessment bodies for assistance in the process. The most established organizations are EPEA in Europe and MBDC in North America, also because these two have been part of the certification scheme since its beginning and were proprietary partners until the accreditation was granted to new partners. Together with the applying company, these assessment bodies work out processes and approaches to achieve the targeted certification criteria. They also play an important role in the onboarding of value chain partners, especially the supplier network. Often the information required for the certification is part of confidential information, like recipes or process details, which suppliers are not willing to share. The accredited assessment body, such as EPEA, works with non-disclosure agreements and serves as an intermediary, thus facilitating the access to required information (Toxopeus et al., 2015). Moreover, the assessment body works with numerous partners and subject matter experts, so a company in the certification process can benefit from the network and the knowledge base (Cradle to Cradle Products Innovation Institute, 2014).

Even though the C2C Products Innovation Institute is presented as a non-for-profit organization, there is a cost component for companies applying for the certificate. As the total costs depend on various factors, e. g. the complexity and duration of the consulting efforts of the assessment body, it is difficult to indicate the total certification cost for a company. A fee schedule published by the C2C Products Innovation Institute indicates the rates for a certification application of a product or product group (3.150 Euro) and the re-certification (1.750 Euro). These don't include the fees that are charged by the assigned assessment body (Cradle to Cradle Products Innovation Institute, 2019c).

2.3.4 Challenges of C2C implementation

The echo of the C2C concept can be described as dichotomous. On the one hand, the idea is celebrated for evincing new ideas and fostering creativity, on the other hand its feasibility on a larger scale is doubted and the certification criteria criticized. C2C and the idea of eco-effectiveness, as the opposition to the concept of eco-efficiency, urges a substantial mindset shift, generating new ideas and approaches towards sustainable innovation. In particular, the idea of creating a positive impact instead of reducing negative externalities can inspire the traditional view on sustainable innovation in research and practice and spur innovation (Bakker et al., 2010). Similar to the discussed business imperative of moving towards a Circular Economy, the C2C concepts presents a possible way of operationalizing the CE concept on the firm-level and can be perceived as an important step in tackling resource problems (Ghisellini et al., 2016; Lieder and Rashid, 2016).

Despite the inspirational character and the innovation potential, the methodological rigour of the C2C concept is also questioned amongst research scholars (Bakker et al., 2010; Reay et al., 2011; Toxopeus et al., 2015). While the concentration on material health raised awareness of product composites and toxic substances (Bakker et al. 2010), the high aspiration to operate with pure organic and benign technical materials might be difficult to realize in practice. Firstly, the nature of biological nutrients is not beneficial per se but rather depends on the amount and its occurrence (Bjørn and Hauschild, 2013). The idea to return biological materials to the biological systems can inhere risks as the impact of such a return cannot always be anticipated and species can show unexpected reactions to certain concentrations (Reijnders, 2008; Reay et al., 2011). Secondly, the strict requirements for the technical materials can pose a challenge as well. Many products consist of so called material composites to ensure certain characteristics, e. g. lightweight design in the automotive industry. The strict requirement to avoid such composites can hinder innovation due to knowledge deficits how to maintain functionality and performance with single technical materials. Besides, the complete removal of contaminating material from the material to be recycled can be very energy-intense (Bjørn and Hauschild, 2013). This lack of attention to the energy use is one other main criticism in the C2C context. As described in chapter 2.3.2, one key idea of the concept emphasizes the necessity of using renewable energy. This requirement goes as far as to largely neglect energy consumption related to production processes. Building on the idea of using only renewable energy, the C2C concept leaves energy efficiency considerations mainly unconsidered in the certification process. In the long run, this might be a realistic option, however in the current transformation phase of manufacturing towards more sustainable processes and products, this can lead to confusion, particularly when applying more established assessments like the Life Cycle Analysis (Bakker et al., 2010; Bjørn and Hauschild, 2013). Also owing to the currently limited availability of C2C conform materials and sufficient knowledge on potential substitutes, the innovation process for C2C products can lead to a longer development time and increased design efforts (Rossi et al., 2006). Additionally, a widely discussed drawback is the scalability and large-scale feasibility of the C2C concept. The aim to fully recycle materials is hindered by the lack of an adequate infrastructure and constraints regarding available market volumes (Bjørn and Hauschild, 2013) although this might also be connected to the currently limited C2C diffusion. Moreover, the idea of a continuous quality preservation during recycling processes still needs to prove its practicability as experts presume a loss of quantity and quality due to corrosion processes (Reay et al., 2011). The constraints also apply to the accessibility of sufficient amounts of biological and technical materials that would be needed if C2C should become a widely established product design standard (Bjørn and Hauschild, 2013).

At the same time, the certification scheme and the underlying conditions are subject to constant adaptations and further development based on the growing experience with certifying companies. Hence, even though the C2C certification program indicates some areas for further improvement, it provides a substantial possibility for companies to operationalize and implement the CE principles. Further, its suitability calls for an accurate and differentiated evaluation based on the products concerned and the respective company context.[24]

2.4 Interim summary and derivation of methodological approach

Building on the idea of a 'quantitative discovery' for abductive reasoning by Bamberger and Ang (2016), the previous chapter described the phenomenological background of the research focus and defined the research opportunity, which shall be explored in the course of the empirical analysis.

While the discussion on sustainability is not new as such, the related research field of Circular Economy and underlying schools of thought, like C2C, seem to have thrived based on current developments in economy and society. Connected research areas, e.g. in the area of environmental management systems (e.g. Hansen and Schaltegger, 2016) or other eco-labels (e.g. Christmann and Taylor, 2006), have investigated motivations for implementation and the process of the adoption or internalization of a new concept. However, a research deficit on studies that quantitatively analyse the implementation success of a CE-related method, such as C2C, remains. Hence, according to the criteria developed by Edmondson and McManus (2007), the state of the research field can be classified as *intermediate*. This stage lies between nascent and mature state of prior theory research and primarily focuses on the exploration of new propositions and relationships based on the analysis of established and new constructs. As the state of prior research is a critical factor when selecting an appro-

---

24 One example for the controversial discussion and complexity of C2C criteria was provided in the context of a publication by Llorach-Massana et al. (2015) who assessed, in juxtaposition with the Life-Cycle Assessment approach, in how far C2C certification predicates the environmental performance of products. The conclusion that the certificate is not always an appropriate indication for environmental products prompted a critical response letter by the C2C Products Innovation Institute (Kausch and Klosterhaus, 2015) who criticized the applied methodology and underlying assumptions.

priate methodology to achieve the research objective, the suggested hybrid approach is applied in the following analysis and the method of choice consists of a qualitative and quantitative part (Edmondson and McManus, 2007). In a first step, a qualitative analysis aims at identifying the most critical factors for companies when initiating and implementing the C2C standard (chapter 3). In a second step, these factors are quantitatively explored with regard to their effects for the implementation success on the organizational level (chapter 4).

Due to the limited empirical and theoretical evidence in the research area of CE and C2C implementation, the suggestion of Bamberger and Ang (2016) to "use a more creative and open approach, drawing hunches from alternative domains and fields" (Bamberger and Ang, 2016, p. 4) has been applied. The exploratory character of the approach allows for an open and accurate investigation of factors and dimensions which are difficult to anticipate beforehand. Hence, the insights generated in the course of the qualitative study (chapter 3) are complemented by extant theory from connected research streams, which proved to be a valuable source of ideas and potential constructs for the quantitative analysis (chapter 4).

The C2C concept has been selected as empirical field for this analysis for various reasons. First, the C2C program targets all kinds of companies worldwide which ensures a certain generalizability of the results. Second, the C2C certification program (see chapter 2.3.3) allows for a detailed analysis of the implementation process as it builds on a progressive certification scheme (basic, bronze, silver, gold, and platinum), which enables companies to individually adjust adoption efforts while at the same time remaining comparable due to the logic of certification levels. Further, the certificate is issued on a product-level, in contrast to many of the more established certifications, e.g. the ISO 14000 or the EU ecolabel, which address a whole company and its processes instead of single products[25]. Thus, selecting C2C certified companies permits an accurate investigation of the processes that are linked to the implementation of C2C innovations and a deeper analysis of different approaches towards implementation. Third, given the fragmented landscape of definitions and terms in the context of Circular Economy, the certificate presents a suitable means of ensuring a common understanding when launching a survey across numerous companies. Hence, the C2C certificate is highly suited for the purpose of better understanding why companies initiate and how they manage the implementation of innovations in the Circular Economy context. In summary, the benefits of selecting the

---

25 The respective websites of the ISO 14000 and the EU ecolabel, being two widely established environmental labels, provide detailed information on the tools, requirements and the certification process; see https://www.iso.org/iso-14001-environmental-management.html for more details on the ISO 14000 family and https://ec.europa.eu/environment/ecolabel/index_en.htm for the EU ecolabel.

C2C concept as an empirical field for this analysis outweigh the disadvantages by far, especially against the background of missing appropriate alternatives in a relatively new research field.

3 Qualitative exploration of companies' experience with C2C implementation

Building on the aforementioned categorisation of the research field, the first step of the empirical analysis consists of a qualitative exploration of companies' motivations and experiences with the C2C certification with the help of an online survey. In the following, chapter 3.1 presents the study design by elaborating on the interview details and the selection of its addressees. Then, an overview of the coding process and main results is presented (chapter 3.2). The chapter closes with the development of propositions, which will feed into the second, quantitative study (chapter 3.3).

3.1 Study design

Despite the growing prominence of the Circular Economy idea, particularly among practitioners, there was no empirical evidence on why and how companies pursue CE efforts (Tollin and Vej, 2012). Given this lack of detailed analyses of CE on the firm-level, the exploratory analysis of this research field begins in a qualitative manner and will be followed up by a quantitative analysis building on the qualitative results (Creswell, 2009; Edmondson and McManus, 2007). As described in chapter 2.3, the C2C standard presents the empirical field for this investigation. In the first phase, interviews were the selected data collection method to be able to grasp the main motivations driving companies to implement the C2C concept as well as their implementation experiences. Since the state of the research field was identified as intermediate, this approach allows to propose new constructs or assume provisional relationships at the end of the qualitative study. It also allows the exploratory testing of new constructs which can be based on extant literature (Bamberger and Ang, 2016; Edmondson and McManus, 2007).

Following the suggestion of a hybrid data collection approach (Edmondson and McManus, 2007), the preceding theoretical problematisation is followed by a qualitative exploration of motivations for C2C implementation and the company's experience throughout the process. Interviews present a suitable method of data collection in this context (Edmondson and McManus, 2007) and will provide the basis for the quantitative survey (chapter 4). The ensuing sections elaborate on the study design by describing the interviewing approach and the data analysis methods. The final section summarizes the preliminary findings by developing propositions which build the basis for the subsequent data collection method and its quantitative exploration.

3.1.1 Method and interview structure

Given the exploratory character of the study, interviews were selected as the qualitative method of data collection. Interviews are particularly suited for the present context as they allow an in-depth investigation of organizational behaviour and at the same time provide the opportunity to explore topics that might be left unseen with other data collection methods (Punch, 2014; Tracy, 2013). Moreover, interview partners present a highly valuable source "for discovery of new concepts rather than affirmation of existing concepts" (Gioia et al., 2013, p. 17). Thus, it is of particular value to involve these 'knowledgeable agents' in the early data collection phases (Gioia et al., 2013). For this purpose, the interview guidelines were designed in a semi-structured nature (Punch, 2014). On the one hand, the flexibility of such an interview form allows the adjustment of the questions to company-specific conditions, for instance depending on the previous experience in the field of Circular Economy, geographical characteristics or customer perception (B2B vs. B2C orientation). Also, it is critical for this kind of study to be able to adjust questions in the course of the interview series in case of important new insights or developments, in particular if the interviewer won't be able to conduct more than one interview with the respective interview partner (Bernard, 2013; Gioia et al., 2013). On the other hand, the structured nature facilitates a comparative analysis of the responses (Punch, 2014). Thus, the semi-structured form is well suited for the intended research of a company's motivations for the C2C certification and their experiences with it.

Qualitative research recommends that oral interviews should not exceed a certain length in order to keep attention and concentration at a consistently high level, both for the interviewer as well as the interviewee (Bernard, 2013). Furthermore, to achieve the highest possible quality of answers, the questions were formulated on the basis of principles of good interview conduct, e. g. simple and unambiguous, neutral and straightforward (Bernard, 2013; Iarossi, 2006; Tracy, 2013). Accordingly, the interview guideline consisted of eight to ten questions, always starting with a short introduction of the interview partners. To ensure comparability between the responses, the opening question asked for the company's understanding of C2C. The following four main sections covered i) motivation or triggers for C2C implementation and certification, ii) internal implementation experience, e. g. specific enablers or barriers, iii) external influencing factors, e. g. supplier management or customer satisfaction and iv) a conclusive assessment, e. g. financial benefits resulting from the certification. Every interview was concluded with the inquiry for own comments. As the main purpose of the qualitative study was to answer 'why' and 'how' questions, the questions were framed accordingly and open-ended peters (Fink, 2013; Peterson, 2000). Furthermore, the interviews themselves were conducted according to recommended interviewing guidelines, such as effective probing to stimulate the respondents to provide more detailed information (Bernard, 2013). According to the nature of semi-structured interviews, the total number of questions varied slightly. On the one hand, the number and exact phrasing of the question was adapted to the respective company background information, which was available upfront from secondary data sources, e. g. whether the company already had extensive experience and a sustainable reputation before the C2C certification. In such a case a question was added to ask for the additional value of the C2C certification besides already existing certifications and programs. On the other hand, the final number of questions could be derived from the initial questionnaire, as 'yes' or 'no' answers from respondents were followed up by more detailed 'why' or 'why not' questions (Peterson, 2000).

3.1.2 Sample selection

Since the data on C2C certified companies' implementation experience was limited, a nonprobability sampling approach (or purposeful sampling) was applied in order to collect the most suitable data to be able to develop a broad overview of the current status in the C2C landscape (Bernard, 2013; Tracy, 2013). Building on the exploratory character of the study, the generated insights should draw a diversified picture of why companies initiated and how they experienced C2C implementation. For this reason, 18 potential interview partners were selected across different industries, countries, targeting B2B as well as B2C customers, holding different levels of certificates (bronze, silver, gold) and ranging from medium to large size, leading to a maximum variation sample (Tracy, 2013). Companies with information on their C2C activities accessible through reports or the corporate websites promised to be a suitable starting point as their external communication suggested a certain openness to share experiences on the topic. Where possible, the direct C2C contact was approached via e-mail. In the case of no explicit point of contact for C2C, the general contact form was used to ask for the right interview partner.

There is no academic consensus about the necessary number of interviews to guarantee a certain theoretical saturation and potential indications range from five up to 50 interviews (Saldana, 2013; Tracy, 2013). Thus, the decision on how many interview partners should be approached also depends on the research objective (Punch, 2014; Sekaran and Bougie, 2010). In the present work, the main purpose of the interviews was an initial exploration of a company's drivers to adopt C2C certification and the related implementation experiences. To be able to grasp a variety of potential factors, it was important to include different kinds of companies in the interview set. After the first C2C companies confirmed the interview request, the selection of further potential interview partners was thus adapted that a wide range of company types could be covered in the study (Bernard, 2013). The first set of interviews was conducted between March and May 2014, the second part in March and April 2015, which yielded an ultimate number of 11 interviews with C2C-certified companies (see Table 1). Since it became apparent that after half of the interviews the mentioned factors became mostly redundant and the number of new insights per interview largely diminished, the data collection was terminated and a certain saturation of knowledge assumed (Sekaran and Bougie, 2010).

3.2 Analysing interview results

The set of 11 interviews built the basis for the initial and exploratory data analysis. Of the 11 interviews, two were submitted in a written form, eight were conducted via telephone and one interview partner suggested doing the interview in person as he was working in Hamburg. The interview language was German or English, depending on the interviewee's background and preference. All oral interviews were recorded with the consent of the interview partners (Bernard, 2013).

The interview length ranged between 15 and 75 minutes, 35 minutes on average and the in-person-interview being by far the longest one. An overview of the interviewees' roles and the company background are summarized in Table 1. After transcribing all interviews with the help of the f4 software (Tracy, 2013), the analytic program MAXQDA 11 was used to analyse the qualitative data.

Table 1: Overview of interviews

ID	Industry	Customer approach	Interview partner's position	Interview type	Interview language	Interview duration (min)
B1	Building	B2C	External Communication	Written	German	N/A
B2	Building	B2B	Sustainability and Product Management	Telephone	English	42
CG1	Consumer goods	B2B/B2C	Head Sustainability Management	Telephone	German	25
CG2	Consumer goods	B2B/B2C	Head of R&D	Telephone	German	42
CG3	Consumer goods	B2C	Founder & CEO	Telephone	English	28
FL1	Flooring	B2B/B2C	Business Development Innovative Solutions	Written	German	N/A
FU1	Furniture	B2B	Chief Engineer Sustainability	Telephone	English	15
PP1	Print & paper	B2B	Technical Marketing Manager	Telephone	German	22
PP2	Print & paper	B2B	National Account Manager	In person	German	75
T1	Textiles	B2B	Senior Expert R&D	Telephone	German	40
T2	Textiles	B2B/B2C	Director Sustainability	Telephone	English	28

The explorative character of the questionnaire and the diverse backgrounds of the selected companies were reflected in the multifaceted statements. This variability of answers was captured with the help of a construct table (Miles et al., 2014), a tool which facilitates analysing the range of a variable's properties as they were collected in the interviews. The main findings with respect to the four main parts of the interviews are presented in the following[26].

---

26 Direct quotes from German interviews were translated into English and company-related information anonymized where necessary to be able to include all interview data in the following analyses.

Motivations for C2C implementation and certification

When asked for the main motivations, the companies revealed different trigger points, yet, two patterns recurred in almost all of the interviews. On the one hand, the implementation was initiated due to a personal connection of a decision-maker, such as the CEO or the company owner, with one of the two C2C founders, e. g. after reading their books on C2C or having heard a talk. As one of interviewee recalled: *"To avoid greenwashing, we were looking for the best standard. And Bill McDonough's book had just come out, 'Cradle to Cradle', and I read the book and I called him"* (interview CG3). In another case, the leadership even invited Michael Braungart to present the C2C idea to the whole company during a strategic planning event in order to onboard the employees *"He [the CEO] heard about Prof. Braungart and invited him"* (interview CG2).

On the other hand, many interview partners emphasized their long history in different sustainability endeavours, so that the C2C concept presented a particularly suitable opportunity to confirm extant activities, as stated by one interviewee *"we have always had a very strong affinity with environmental issues [...] it's a key element of our business strategy"* (interview T1). In some cases the C2C concept pushed already established sustainability efforts even further and therefore was selected, like one interviewee explained *"we have recognized that for us this [C2C] means a further development [...] we consider this to be an innovation of the eco-labels and a way to look beyond our own horizons"* (interview PP1).

In a few cases, the C2C certification presented a rather evident step as the products with which the companies applied for the certification already fulfilled all criteria and didn't require any change. As one of the interviewees elaborated *"for the first certification [...], we presented this product [...] and it was directly certified without any modification of the product"* (interview B2).

However, some of the mentioned motives were also mainly business-oriented and related to the competitive situation of the companies. E. g. *"for our products this is a unique selling proposition, which has not yet been implemented by our competitors"* (interview PP1) or as in the case of one of the building companies *"C2C was a commercial tool to sell the product on the market, because we were the first to get that kind of certification in our field of application"* (interview B2). Reflecting this variety of motives, in most of the cases, the motivation was grounded on a combination of the factors, as one interviewee summarized their decision after the CEO had discussed the C2C potential of their products with Michael Braungart: *"through the conversation with Michael Braungart we have been shown ways to become even better, i. e. by abandoning the use of chemicals or making our own electricity through heat recovery. So we are always looking to become one step better, since we know that we are in a highly competitive market"* (interview PP2).

Internal implementation experience

The part of the interview on the internal implementation experience consisted of questions concerning the C2C impact on the innovation process and internal hurdles or enablers. The emerging topics covered various challenges and benefits and highlighted a certain complexity concerning the involvement of many different actors within the company. On the one hand, some interviewees showed a clear frustration, as for instance one interviewee reported *"when you have to do everything yourself, because the respective division managers have ignored the whole thing [...] it makes you feel like a fool"* (interview CG2). Others described conflicts with other departments which were affected by necessary changes in the course of C2C implementation as well: *"to convince our in-house marketing department to go for such colours, which are in line with the C2C principle. That was relatively difficult and also tedious."* (interview CG1). On the other hand, interviewees also observed committed colleagues or related departments that contributed to a more successful experience, e.g. *"That has kind of accelerated innovation in our company and uniqueness of the way we build [our products] since the time we consistently applied the C2C framework to all of our new product designs."* (interview FU1).

One rather controversial topic was related to the cost and effort of C2C implementation and its (re-)certification. In many interviews, the interviewees expressed a multifaceted trade-off between resulting advantages and a rather critical perspective towards the certification process and the related costs, e.g. *"the whole implementation of Cradle to Cradle increased our quality but to be honest it is a very difficult process because you are going beyond legislation, beyond everything and this is something a company needs to really invest in"* (interview T2) or *"this [C2C] truly gives an important though-provoking impulse, but it is certainly not available at zero cost"* (interview PP1). Yet, there were also cases of a more explicit critique concerning the certification, e.g.: *"the currently offered certification procedures are unfortunately much too expensive for small-size or private buildings"* (interview B1).

Others described that C2C implementation led to improved cost and risk management on the company-level: *"The use of recycled waste as well as the introduction of closed cooling water circuits or energy savings due to energetic measures allows us to use these savings as an investment in new materials and substances"* (interview FL1). Yet, it was emphasized in several cases that the process of C2C implementation needs capacities and capabilities, especially in Research and Development. Two interviewees recollected: *"I mean, you suddenly have to put aside 50 years of research and development. Well, it's not so easy"* (interview CG 2) or *"But unfortunately, it's not so easy to reinvent physics and chemistry [...] that's rather a tough uphill struggle"* (interview PP1). Further, a proper integration within the innovation process was mentioned, e.g. *"it is really integrated within the design process. Our engineers are all trained on the concept of C2C thinking and we have a sustainability team that is part of the design and engineering team for all new products*" (interview FU1) or as described by the interviewee in the case of T1: "*in fact, C2C philosophy must be considered early in the development of a product already [...] otherwise, if you try to squeeze it in with hindsight, it will not be affordable and the customer will not be inclined to buy more or to buy something new just because of C2C*".

External influencing factors

The discussion on influencing factors from the external company environment was dominated by comments on the collaboration with the certification-related organizations and the complexity of the supply chain management. Also here, benefits and challenges needed to be balanced out against each other. On the one hand, interview partners reported difficulties with respect to sourcing and procurement of C2C conforming materials, e. g. concerning necessary substitutes for excluded components. Sometimes this was a matter of available quantities, such as described by one interviewee "*therefore, it is simply a question of the waste volume available on the market that makes it very difficult or impossible. [...] We have basically reached the limits of what the industry can do.*" (interview CG2).

In other cases, the challenge resulted from incapable or reluctant suppliers as pointed out by the head of sustainability management (CG1): "*it was always a bit of a question mark, what and how our suppliers work [...] because the recipes [...] are the know-how of the supplier and they don't openly give them out. [...] So, the main difficulty was actually to get our suppliers to build up such a trust with the certifier, in this case the EPEA, with our help, that they would hand over the recipes.*" On the other hand, the requested focus on the supply of C2C suitable materials also presented a source for quality improvement in some cases, as explained by one interviewee: "*there is a much more intimate handle on the material received, so due to the higher level of oversight on the suppliers, they are motivated to supply more consistent and higher quality material*" (interview FU1).

In the supply chain context, the critical role of EPEA or MBDC, which was pointed out in all interviews, became quickly evident. At the same time, the certification process itself, in which, at the time of the interviews, EPEA was playing a central role in consulting, assessing and certifying products[27] was also a subject of all interviews. Besides the supporting role of EPEA/MBDC, such as described in the context of supplier collaboration, it was also mentioned that the certification process was not always satisfactory, e. g. "*so we had a long wait for our [product], even though we had made [it] according to the rules that were given to us. It took an eternity with*

---

27 See chapter 2.2.3 on the role of EPEA and MBDC as well as the details of the certification process.

EPEA [...]. We had the [products] on the market, but we could never say that they were C2C certified" (interview PP2).

Another perspective on the external influencing factors pointed at the critical role of the consumers and the economic situation. One interviewee, for instance, described the effects of the economic crisis in 2008: *"[...] but one recession and it's done. So C2C has really suffered I think through that, because customers are exhausted, consumer just want a low cost product"* (interview CG3). This statement was in line with several observations from the interviewees concerning the role of the customer, the customer's willingness to adjust consumer behaviour or the willingness to pay. As described in some cases, the implementation of C2C would also require a changed consumption pattern, e.g. not discarding old products randomly but returning it to the producer disposal of used products. This is more straightforward for some products than for others, e.g. as one interviewee explained: *"You have to organize the process so that the customer has minimal work [...]. If then, let's say, the whole stuff should be sent somewhere [...], transporting 15 kilos to the central collection point, maybe even paying money for sending it, nobody does that."* (interview T1).

The collection problem can also apply in the B2B environment since the responsibility then lies with the central procurement departments, which might be reluctant to changing routines or evaluation processes, as exemplified in the interview with CG2: *"Most of our customers are large firms [...]. The purchasing department then buys the [product] centrally and they just look, what they bought last year and buy the same next year."*

The potential causes for these effects of external influencing factors were explained with different approaches. When customers are already sensitized to CE-related topics and informed about the concept of C2C, it seemed easier to convey the benefits and convince them of the C2C certified products. In other cases, several interviewees emphasized the complexity of C2C which breaks with some of the prevalent sustainability-related habits, hence *"it required lots of explanation"* (interview PP1) and *"to impart the topic C2C to the end user is of course also proved to be difficult"* (interview CG1).

Conclusive assessment

Ultimately, the conclusive feedback of the interview partners was multi-faceted ranging from optimistic and satisfied to a rather frustrated and resigned echo, with some expressing a willingness to continue and expand the C2C certification program and others planning to abandon it in the near future. In most of the cases, however, the final assessment remains ambivalent and interview partners express a careful weighing of benefits and cost. For instance, like in the case of CG3, the interview partner holds on to the belief in the C2C concept and the fit to his company *"we love it as a story telling mechanism"*. At the same time, he expresses also a certain disappointment: *"But my expectations 10 years ago was that they would be a real PR engine, that they would get out there and tell the story of cradle to cradle with all they have and I think that they have done a very poor job."* (interview CG3).

This ambiguity was observed for most of the companies. Almost all interview partners reinforced that they believe in the C2C concept, however in most of the cases articulated some disappointment with respect to different topics. For some, the customer response was a critical point, e. g. *"But the point is, the passivity of customers is just massive."* (interview CG2). For others, the question of the potential market remained difficult to assess, e. g. *"actually, all we needed to commercialize [the product] was already there. What has not been there is the market in Germany."* (interview PP2). In the more critical discussions, the question of the further development and particularly the time horizon, remained mostly unanswered, as pointed out by the interview partner from T2: *"Our customers are still thinking linear. So, we still are in a transition stage"*. The uncertainty also applied to potential expansion possibilities *"For the next 20 years, I don't see any prospect of C2C giving us an advantage in these markets"* (interview T1). Of the interviewed companies, three confirmed the abandonment of C2C certification, two were undecided and the remaining six companies planned to continue.

Coding process

Based on the building blocks, the further coding process allowed for the identification of underlying patterns and more detailed classification of the topic. Since the theoretical basis was limited, the primary aim of the systematic content analysis process was to identify the most relevant motivational and organizational factors in the context of C2C implementation in a first step. To ensure rigour of the study while "still retaining the creative, revelatory potential for generating new concepts and ideas" (Gioia et al., 2013, p. 15), the approach of identifying new insights followed the work of Gioia et al. (2013). This approach, although adaptable to the respective research objective and context, is an appropriate means for the identification of new concepts, which then can be developed into new constructs (Gioia et al., 2013). It suggests an initial coding process, followed by a consolidation of these initial codes (1st-order terms) which are then organized into themes (2nd-order themes). These build the basis for an overarching dimension. Ultimately, the terms, themes and dimensions are synthesized in an illustrative 'data structure' (Gioia et al., 2013). Consequently, this systematic process is highly suited for the study at hand.

The coding of the eleven interviews was conducted following the described approach and resulted in over 340 initial codes, nine 2nd-order themes and three aggregate dimensions (see Figure 8). Given the diversified set of the selected interview partners, it was little surprising that codes revealed very different approaches toward the C2C implementation. As the primary goal of the study was to create a first understanding of why and how companies engage in C2C, counts or rankings of codes were not part of the analysis. At the same time, for some questions, the findings were more uniform than expected, e.g. when synthesizing the codes about the customer feedback which was largely dominated by disappointing results from the companies' perspectives. With the emergence of the 2nd-order themes, the iterations between the data and existing relevant literature were intensified (Gioia et al., 2013) and reflected the scarce theoretical basis in the area of C2C implementation on the company-level. While the relevant literature covers a range of motivations for sustainability-oriented organizational activities (e.g. Bansal and Roth, 2000; Hahn et al., 2015), it transpired that a more holistic theoretical perspective on motivational drivers and the following implementation process of a CE-related approach, such as C2C, was not present up until then.

To be able to synthesize the findings as a basis for the proposition development and the subsequent quantitative analysis, the 2nd-order themes were framed in a neutral way. This was encouraged through interviews, in which partners complemented narratives on rather negative implementation experiences with ideas on what would have helped to overcome these barriers. Thus, it would be possible to also include these factors in the survey. As the interview partners were always asked at the end of the questionnaire whether they would like to continue pursuing the C2C implementation and why (not), the data structure was completed with codes that described the satisfaction of the companies.

Figure 8: Data structure of qualitative results

3.3 Deriving propositions for further analysis

The discovery-oriented coding process allowed the identification of four groups of motivators and a differentiation between three categories of contextual factors that have influence on a company's C2C implementation process (see Figure 8). These results built the basis for the following quantitative data collection with the themes and dimensions presenting cornerstones of the questionnaire, complemented by theoretical constructs where available and applicable. The next step comprised the proposition development and hereby followed the recommendations for abductive reasoning which "calls for propositions to be inferred and evaluated on the basis of empirical findings" (Bamberger and Ang, 2016, p. 4). By summarizing and reflecting the main results, propositions are a suitable means to present study conclusions as a "coherent set of explanations" (Miles and Huberman, 1994, p. 75). The developed propositions can then be tested further with the help of other data collection methods, such as quantitative surveys, and by integrating extant literature (Miles and Huberman, 1994). Particularly, in the present exploratory research design, propositions can present a suitable means to derive first predictive proposals based on the first findings and at the same time anticipate 'if-then' or 'why-because' relationships (Miles et al., 2014).

The companies' conclusive assessment and intent to prolong, expand or abandon C2C was a result of balancing various influencing factors that occurred along the implementation and certification process. The weighing up led to an eventual perception whether the implementation and certification can be regarded as a success or not. Based on the data structure, it appeared that the major influences on this final assessment stemmed from three different factors:

1. whether expectations were met,
2. the organizational barriers encountered during implementation and
3. the customer feedback.

Thus, the first proposition summarizes these potential influencing factors as follows:

P1: A company's decision whether to further pursue C2C depends on the satisfaction with the outcomes. In turn, the satisfaction is influenced by a company-specific and differentiated balancing of various factors. These factors include initial motives to engage with C2C, the organizational processes (enablers or barriers) involved, and the final customer feedback.

Since the interviews started with openly asking for motivations for implementing C2C and applying for C2C certification, the identified groups of motives present first indications for drivers. Besides already researched motivators for the initiation of such an implementation effort, e.g. seeking for legitimation (Bansal and Clelland, 2004; Berry and Rondinelli, 1998) or seeking for a competitive advantage (Crilly et al., 2012; Esty and Porter, 1998), the analysis also revealed that some companies perceived the C2C concept and its characteristics as particularly well-matching to company values or goals. These companies were more experienced in sustainability efforts and in most cases had already developed sustainable innovations, sometimes also including other certification programmes. Further, the active and empowering role of the executive board or company owner proved to be a promising trigger for engaging in C2C. A univocal relationship between motives and the eventual satisfaction could not be identified in the course of the interview series. Thus, two propositions are developed in order to better grasp a clearer pattern of motives and their role in the final assessment: P2 reflects the main motivations to implement C2C and apply for a certification, and P3 addresses their effect on satisfaction.

P2: The decision to engage with C2C can be triggered by different motivations, which range between more pro-active and opportunity-related motives to more reactive and threat-avoiding motives. Further, initiation can also emanate from decision-makers, often being aligned to the corporate strategy or goals. One single dominant trigger can hardly be isolated within this group of motives.

P3: The company's motivations, which are most often expressed in terms of respective expectations, have an influence on the company's satisfaction with the C2C implementation and certification process.

Alongside the motivations, the interviews revealed that companies went through positive and negative experiences regarding their internal and external environment. The importance of various company-specific factors during the adoption of new practices has already been the subject of numerous research studies, even though their relevance in a sustainability-related context was still rather scarce. In their analysis on why companies 'go green', Bansal and Roth (2000) identified three contextual dimensions that affect the motivations. 'Issue salience' referred to the criticality of ecological issues to different groups of the organization, 'field cohesion' captured network ties and relationships between internal and external stakeholders of a company, and 'individual concern' addressed the individual attitudes of the organizational members towards ecological issues, such as personal values or beliefs. The effects of this contextual factors also depended on the characteristics, or profiles, of the companies, and the authors developed propositions for further testing (Bansal and Roth, 2000). Further, specific skills and capabilities have been pointed out as critical influencing factors, particularly in the context of sustainable innovations which often necessitate new knowledge or adaptations of prevalent R&D practices (Ketata et al., 2015). In line with several interview results, the deviation from current practices and existing knowledge confronts companies with major challenges, for some more than others, depending on their experiences with sustainability-related practices, e.g. with respect to innovation methods or supplier networks (Seebode et al., 2012). The shift towards sustainable innovation might also require a shift of perspective concerning return on investments (Chappin et al., 2015), which was also emphasized in all of the interviewed companies. Based on the findings from previous research and gathered observations the next proposition resulted in two sub-points as follows:

P4a: The adoption of the C2C certification is influenced by the organizational context in such way that internal and external stakeholders are critical for a successful implementation.

P4b: Current processes as well as existing skills and capabilities need to be reviewed and potentially adapted in order to achieve satisfactory results.

The organizational context is an essential influencing factor for how the C2C adoption can be realized. The interviews pointed at several enabling or impeding circumstances that stemmed from the organizational set up, i.e. the organizational complexity and its effects on C2C adoption. This would for instance include various levels of responsibilities for C2C-related matters that need to be aligned. In order to be able to account for the large variance of corporate contexts, the following proposition is framed:

P5: The effect of organizational context factors on the company's satisfaction with C2C implementation and certification also depends on the prevailing organizational complexity which influences C2C adoption within the company.

A less ambivalent effect on a company's satisfaction could be observed with regard to how the C2C product was perceived by the market. In some cases, even though the company itself seemed decided to continue the C2C path regardless of organizational hurdles or additional cost, a negative customer feedback could thwart these plans. On the contrary, some interviewees also emphasized that the positive customer perception was a decisive factor to continue, in some cases despite a more critical view on C2C from within the company. Based on this observation, the following proposition was developed:

P6: In contrast to a more equivocal relationship between motivational and organizational factors with the company satisfaction, the market success of the C2C certified product(s) clearly positively influences the eventual satisfaction with C2C implementation and certification.

Since the qualitative study did not allow for drawing early conclusions on the correlations between presumed relationships described in the propositions, the analysis is followed by a subsequent quantitative study in the following chapter.

4 Quantitative exploration of motivational factors and organizational enablers

The following section introduces the quantitative exploration method, which was selected for the present research context. After elaborating on the study design and the operationalization of the variables included in the analysis (chapter 4.1), the data is investigated thoroughly to prepare for further analysis (chapter 4.2). Chapter 4.3 then descriptively outlines the study results to give an overview of the company and respondent profiles as well as the main survey variables. After testing required validity and reliability criteria, an exploratory factor analysis (chapter 4.4) lays the groundwork for the following multiple regression analysis (chapter 4.5).

4.1 Research design

The quantitative analysis is a suitable means to further understand the described phenomenon and allows for a better generalizability of the observed results, which in turn can provide theoretical contributions, e.g. by allowing for the observation and identification of patterns resulting from the data analysis (Bamberger and Ang, 2016). Even though, quantitative and statistical analyses are more typically conducted in the case of confirmatory research studies, the collection of quantitative data can also provide an appropriate approach for "discovery-oriented research activities and objectives" (Bamberger and Ang, 2016, p. 2). It further allows for the development of new constructs or variables which can help to better understand a fairly unexplored research interest (Bamberger and Ang, 2016).

Furthermore, as indicated earlier, the state of the prior theory can be considered intermediate (Edmondson and McManus, 2007), thus the data analysis method of choice was selected to be hybrid, which also calls for a quantitative analysis based on survey data. Due to the limited existing empirical evidence in this context, an exploratory method is appropriate and applied in the following. So far, quantitative analyses at the company or organizational level have been scarce and research is dominated by anecdotal analyses, also often overlooking concrete discussions at the implementation level[28]. Thus, to expand on the qualitative findings, the quantitative analysis focused on the factors related to the implementation process of the C2C concept, including the foregoing motivations and the organizational conditions for successful implementation and certification (Creswell, 2009).

4.1.1 General research framework and operationalization of variables

To be able to answer the underlying research questions and find evidence for the developed propositions, the subject of analysis is twofold. On the one hand, companies are asked for their motivations for engaging in the C2C certification scheme and for their underlying expectations. On the other hand, the organizational context of C2C implementing companies is explored as well as the company background, e.g. size, age or industry. As revealed in the qualitative study, the motivation to continue, expand or abandon the C2C certification also depends on the company's satisfaction after the implementation, hence on whether the implementation is regarded as successful. Therefore, the perceived satisfaction appears as a critical factor for a company's decision to keep following the C2C pathway or abandon this route again. In turn, this could influence the argumentation why and under which circumstances a shift towards a Circular Economy, here expressed by the implementation of the C2C concept, is more or less promising. Thus, the research framework is developed to explore which motivational factors and which organizational enablers or hurdles impact a company's satisfaction with their C2C activities (Figure 9).

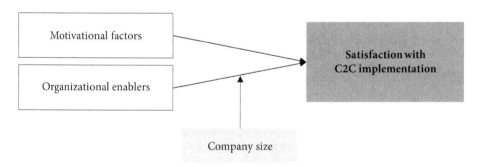

Figure 9: Conceptual research framework[29]

---

28   See chapter 1.2.
29   Author's illustration.

Following the suggestion of Bamberger and Ang (2016) the operationalization of the variables was built on extant literature as well as newly developed measures derived from the qualitative study. Given the exploratory character of the research framework, no *a priori hypotheses* (Bamberger and Ang, 2016) were developed as the aim of the study is to identify relationships and patterns related to the emerging research field of C2C implementation and to ascertain new clusters of factors affecting its organizational success (Bamberger and Ang, 2016; Edmondson and McManus, 2007). Such a quantitative, but more open approach "to describe and examine organizational problems, anomalies, and management-related phenomena lying beneath the radar may serve as critical means of laying the groundwork for theory generation" (Bamberger and Ang, 2016, p. 1).

The selection of the dependent and independent variables was based on empirical evidence from more general New Product Development (NPD) literature and complemented with findings from the qualitative study, which reflects the specific C2C context. For decades, scholars have been researching success factors of product and process innovations and how they can be successfully introduced, hence the list of factors that differentiate innovation success and failure is long (e. g. Cooper and Kleinschmidt, 1987) and contains product-, firm- and customer-related components, such as superior product features, clear project management or technology synergy. Sustainability-related innovations certainly relate to a number of the already researched factors, however based on the additional inclusion of sustainability aspects are subject to a further complexity and might necessitate a different perspective (Ketata et al., 2015).

Dependent variable

Reflecting proposition P1 on the company's retrospective assessment of C2C implementation[30], the perceived benefit of integrating C2C was measured with a construct describing company satisfaction. This construct was based on both the qualitative study results and the empirical work of Chun and Davies (2006). Originally tested in the context of stakeholder satisfaction with a company's brand image, the construct was adapted to reflect a company's satisfaction with the C2C implementation. One of the four original items was excluded as it did not fit the context, instead four additional items were added based on the results of the qualitative study. The construct resulted in seven items that were all measured on a 5-point Likert scale (Table 2).

---

30 See chapter 3.3 for the comprehensive elaboration of the propositions.

Table 2: Survey items on satisfaction with C2C implementation

	Item 'satisfaction'	Scale	Source
S1	We would recommend the implementation of C2C to our business partners	1 = Strongly Disagree, 5 = Strongly Agree	Chun and Davies (2006)
S2	We are pleased to be associated with the C2C standard	1 = Strongly Disagree, 5 = Strongly Agree	Chun and Davies (2006)
S3	The implementation of C2C spurred innovation in the company	1 = Strongly Disagree, 5 = Strongly Agree	Qualitative study
S4	The company's image has improved because of the C2C implementation	1 = Strongly Disagree, 5 = Strongly Agree	Qualitative study
S5*	The cost of C2C implementation is too high compared to the financial benefit	1 = Strongly Disagree, 5 = Strongly Agree	Qualitative study
S6*	The cost of C2C certification is too high compared to the financial benefit	1 = Strongly Disagree, 5 = Strongly Agree	Qualitative study
S7	Please indicate your company's overall satisfaction with the C2C implementation	1 = Very low, 5 = Very high	Chun and Davies (2006)

* Reverse coded items

Independent variables

Understanding why companies engage in sustainability-related activities has already been researched by numerous scholars (e.g. Bansal and Roth, 2000; Crilly et al., 2012). However, as the landscape of potential definitions, tools and methods for sustainability-related innovations is very broad, there is no widely accepted and empirically supported consensus on the motives of companies initiating sustainability standards such as the C2C certificate. Most approaches of classifying motivational factors have in common that scholars identify motives that are related to potential competitive advantage and those that are more related to external pressure, e.g. from shareholders or the government (Bansal and Clelland, 2004; Bansal and Roth, 2000; Crilly et al., 2012; Hansen et al., 2009; Kennedy and Fiss, 2009). These two groups of motives were also reflected in the interviews[31]. While most of the studies have a conceptual character or are developed on single case observations, the research projects of Crilly et al. (2012) as well as Kennedy and Fiss (2009) have introduced and empirically tested constructs which describe corporate motivations for implementation of new concepts. In their research on TQM adoption in US hospitals, Kennedy and Fiss (2009) set out to identify why and how companies implement new practices

---

31 See proposition P2 in chapter 3.3.

by examining the interplay between economic and social considerations in adoption decisions. Even though not directly linked to sustainability, the analysis is well suited as a basis for the present research context as it provides a holistic approach to better understand the process of adopting a new practice in a company. As a second source of theoretical constructs, the work of Crilly et al. (2012) provided survey items that were originally investigated under the theoretical lens of decoupling policy and practice in response to institutional pressure. The authors used the example of CSR practices, which also pointed at the potential contrast between responding to external pressure by securing legitimacy and a more intrinsic motivation based on a potential amelioration of the competitive position. The study also underlined the importance of initial motivations for a later communication and perception of the selected CSR efforts (Crilly et al., 2012). The work of Bansal and Roth (2000) was used to check for compliance-related motivations. To complement the list of motivational factors extracted from theory, the motivations identified in the qualitative analysis were added (see Figure 8 for data structure). In this context, the exploration of motivations is not only relevant to identify most prominent motives for C2C initiation, but also to better understand their respective influence on the company satisfaction. The interviews have shown that in some cases the company's perception of whether the C2C implementation was successful or not also depends on the starting point, hence the expectations and motivations before pursuing the C2C certification, which refers to proposition P3[32]. All items were modified to the C2C context in their wording. Table 3 shows the final set of motivational factors, all measured with a 5-point Likert scale ranging from 1 = Unimportant to 5 = Very important. The list was closed with an open question in order to capture additional factors.

---

32 See proposition P3 in chapter 3.3.

Table 3: Survey items on motivations to implement

	Items 'motivation'	Source
M1	The fit of the C2C concept with the company's philosophy	Qualitative study
M2	The opportunity to get the company's sustainability efforts approved by an independent certification institute	Qualitative study
M3	The company CEO's / top management's strong endeavor to achieve the C2C certification	Qualitative study
M4	The expectation to contribute to the company's cost reduction	Crilly et al. (2012)
M5	The expectation to contribute to the company's risk reduction	Crilly et al. (2012)
M6	The expectation to increase sales	Crilly et al. (2012)
M7	The expectation of C2C to be a source of new opportunities	Crilly et al. (2012)
M8	The potential loss of market share if the company does not implement C2C standards	Kennedy and Fiss (2009)
M9	The competition from other C2C certified companies	Kennedy and Fiss (2009)
M10	The expectation to improve the quality of the company's product(s)	Kennedy and Fiss (2009)
M11	The expectation to be perceived as a market leader	Kennedy and Fiss (2009)
M12	The expectation to improve customer satisfaction	Kennedy and Fiss (2009)
M13	The demand of the company's customers for C2C	Qualitative study
M14	The expectation to better comply with current legislation	Bansal and Roth (2000)

In addition to the motivational factors, the organizational context proved to critically influence the way companies implement a new practice or managerial concept (Chappin et al., 2015; Rindfleisch and Moorman, 2001; Song and Montoya-Weiss, 2001). This implies, that once a company decides to adopt C2C, enabling and hindering conditions within the organization can have a substantial impact on the course and success of the implementation. Thus, to relate to proposition P4 (chapter 3.3), five groups of potential influencing factors were identified based on previous theoretical work and on the qualitative analysis.

When implementing a new concept or developing a new product, the importance of relationships has often been researched and found critical (Johansson, 2002; Koen et al., 2002; Wheelwright and Clark, 1995). Rindfleisch and Moorman (2001) analysed new product alliances with a particular focus on "joint acquisition and utilization of information and know- how" (Rindfleisch and Moorman, 2001, p. 1).

Furthermore, in the specific context of a certification program, scholars have underlined the importance of alliances and partnerships between the certification-related actors and the adopting companies in previous research on voluntary sustainable standards[33] (e.g. Blind and Mangelsdorf, 2016; Helms et al., 2012). The close interrelation of the assessment bodies EPEA and MBDC to certifying companies was particularly relevant during the time of the data collection as they were the main partners for both, assisting companies in the necessary analyses and activities to achieve the certification as well as granting the final certification mark. This process has been changed in the last years, so that there are different and independent parties responsible for the support of the activities prior to the certification and the final certification grant (see chapter 2.3.3). As this was not the case at the time of the survey, the items are asking for the partnerships with EPEA and MBDC only and do not consider today's C2C PII or other assessors who are authorized as well to support certification applications. The quality of such a partnership, what Rindfleisch and Moorman (2001) describe as 'relational embeddedness' seems to be particularly suitable for the present study as the attainment and management of new knowledge in the course of the C2C certification was one of most frequently mentioned points during the interviews. The wording of the items was adapted to the specific C2C context (see Table 4). A 5-point Likert scale reflected the agreement to the respective statements, ranging from '1 = Strongly disagree' to '5 = Strongly agree'. Also with respect to the relationship, the theoretical items have been complemented with variables that emerged from the qualitative study results.

---

[33] In the recently published study of Smits et al. (2020), a subset of the participating C2C companies has been analysed in-depth to identify the interplay of motivational factors, the experiences during the implementation process and the implementation extensiveness.

Table 4: Survey items on organizational enablers – relationship with certification partner

	Items 'relationship with certification partner'	Source
PAR1	We feel indebted to EPEA and/or MBDC for what they have done for us	Rindfleisch and Moorman (2001)
PAR2	The company's employees share close social relations with the employees from EPEA and/or MBDC	Rindfleisch and Moorman (2001)
PAR3	Our relationship with EPEA and/or MBDC can be defined as "mutually gratifying"	Rindfleisch and Moorman (2001)
PAR4	We expect that we will be working with EPEA and/or MBDC far into the future	Rindfleisch and Moorman (2001)
PAR5	The collaboration with suppliers was extremely important to implement C2C standards	Qualitative study
PAR6	The company extensively collaborated with scientific partners, e. g. universities, in order to implement C2C standards	Qualitative study
PAR7	The company extensively exchanged knowledge with other companies in order to implement C2C standards	Qualitative study

As the qualitative study also indicated that there is a variety of different degrees of implementations across the certified companies, the level of implementation was included in the quantitative analysis. From an organizational perspective, the implementation process itself has been in the focus of scholars when researching the successful adoption of new managerial practices. In their work on the internalization of sustainable practices, Chappin et al. (2015) reflect on the challenges of going from the decision to implement to actually internalizing and adopting a practice throughout the company. Based on this research, the construct on the level of implementation has been adjusted to the C2C context and included in the survey[34]. Due to the scarce empirical basis in the field of Circular Economy implementation, it was drawn upon results from a related research stream, referring to the work of Kennedy and Fiss (2009) on why and how hospitals implement Total Quality Management (TQM). Linking initial motivation to the actual internalization of TQM, they analysed different forms of motivations, also

---

[34] One other important component in the analysis of Chappin et al. (2015) was also the "willingness to cannibalize", reflecting a company's will and ability to divert from existing products or technologies. It "refers to the extent to which a firm is prepared to reduce the actual or potential value of its investments" (Chandy and Tellis, 1998, p. 475). Even though this construct was also included in the survey, the number of responses was not sufficient for the items to be included in the further analyses. Thus, the construct is not further presented in this chapter.

in the light of implementation timing, and their impact on the extent of implementation (Kennedy and Fiss, 2009). Hence, the applied constructs seemed to be a highly suitable basis for the survey at hand. To complement the construct at the implementation level, the theory-based items were reconciled with the data structure from the qualitative study. This led to the addition of one item which intended to reflect the relevance of the involvement of the supply base, a frequently mentioned theme during the interviews which was not yet present in the list of items. Thus, the level of implementation was operationalized with seven items in the survey (see Table 5).

Table 5: Survey items on organizational enablers – level of implementation

	Items 'level of C2C implementation'	Scale	Source
IMP1	The company has integrated the C2C standards in procedures and work instructions	1 = Strongly Disagree, 5 = Strongly Agree	Chappin et al. (2015)
IMP2	The company has identified specific persons and positions responsible for C2C implementation	1 = Strongly Disagree, 5 = Strongly Agree	Chappin et al. (2015)
IMP3	The company keeps records of the training provided to staff in relation to the implementation of C2C standards	1 = Strongly Disagree, 5 = Strongly Agree	Chappin et al. (2015)
IMP4	The company has obliged its supply base to supply according to C2C standards	1 = Strongly Disagree, 5 = Strongly Agree	Qualitative study
IMP5	The company has adapted the C2C implementation procedures to its various business departments, business units or plants/warehouses	1 = Strongly Disagree, 5 = Strongly Agree	Kennedy and Fiss (2009)
IMP6	The company has integrated the C2C standards in its computerized and other administrative systems	1 = Strongly Disagree, 5 = Strongly Agree	Kennedy and Fiss (2009)
IMP7	Please indicate the extent to which you believe that at this point in time C2C philosophy, standards, and methods have been implemented throughout your company	1 = Not at all implemented, 5 = To a great extent implemented	Kennedy and Fiss (2009)

The level of implementation can also be linked to the degree of change of the certified product compared to the existing products, hence the necessary adaptation of the product development processes compared to established processes in place.

Numerous researchers have underlined how critical it is for a new product development project to fit to the company's resources and capabilities (Cooper, 1988; Danneels and Kleinschmidt, 2001). A well suited construct to describe such 'fit with the available R&D, engineering, and production skills and resources' is provided by Song and Montoya-Weiss (2001), who examined different influencing factors on product development activities. The construct, called 'technical synergy', can affect the development process proficiency and product competitive advantage, depending on the company's perceived technological uncertainty (Song and Montoya-Weiss, 2001). Thus, the construct suitably expresses a theme that was present in the interviews when interview partners raised the criticality of adapting to the requirements of the C2C certification which were a stretch for some more than for others. The theoretical construct 'technical synergy' was rounded out with some of the major statements from the qualitative study, resulting in eight items which were measured on a 5-point Likert scale (1 = Strongly Disagree, 5 = Strongly Agree) and are presented in Table 6.

Table 6: Survey items on organizational enablers – technical synergy

	Items 'technical synergy'	Source
TS1	The company's R&D skills were more than adequate for implementing the C2C standards	Song and Montoya-Weiss (2001)
TS2	The company's engineering skills were more than adequate for implementing the C2C standards	Song and Montoya-Weiss (2001)
TS3	The company's R&D resources were more than adequate for implementing C2C standards	Song and Montoya-Weiss (2001)
TS4	The company's engineering resources were more than adequate for implementing the C2C standards	Song and Montoya-Weiss (2001)
TS5	In the course of implementing C2C standards, the company made fundamental changes to the existing product and processes	Qualitative study
TS6	The innovation process had to be significantly adapted to the C2C standards	Qualitative study
TS7	The implementation of C2C standards had a significant impact on the entire value chain of the company	Qualitative study
TS8	The C2C certified products were mainly introduced as a new product line	Qualitative study

Building on the interviews, one further construct was added to the group of organizational enablers, which reflected specific context factors related to the C2C certification so that the anecdotal referencing of certain factors can be empirically tested. This list resulted in four items which were present in most of the interviews. In many cases they were discussed controversially as they were present in some of the companies and absent in others (Table 7).

Table 7: Survey items on organizational enablers – C2C specific context factors

	Item 'C2C specific context factors'	Scale	Source
CC1	The implementation of C2C standards is well incentivized by political regulations	1 = Strongly Disagree, 5 = Strongly Agree	Qualitative study
CC2	The top management provided guidance during the implementation of C2C standards	1 = Strongly Disagree, 5 = Strongly Agree	Qualitative study
CC3	The implementation of C2C standards would be easier if more companies from the industrial sector would implement it as well	1 = Strongly Disagree, 5 = Strongly Agree	Qualitative study
CC4*	Did the company's internal stakeholder (e.g. marketing, product development, sales) show resistance to the implementation of C2C standards?	1 = Never, 5 = Always	Qualitative study

* Reverse coded item

Moderator variable

Building on proposition P5, the analysis of the effects on the implementation of new practices coming from the organizational context also prompts an assessment of different degrees of organizational complexity and how companies manage internal processes in general. In this case, the moderation analysis is an appropriate means as it can uncover a potential dependency of a relationship between the independent and the dependent variable on a third variable – the moderator variable (Dawson, 2014; Hayes, 2013). To examine a potential moderating effect of the organizational complexity, the company size has been used as a proxy. Thus, this variable will be introduced as moderator variable in the following research model and tested for a direct effect on the dependent variable as well as indirect interaction effects (see Figure 10 for a conceptual illustration of the moderation effect).

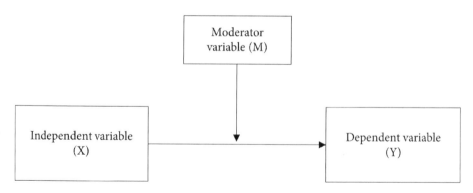

Figure 10: Conceptual diagram of moderator effect[35]

The suitability of this measure has been widely researched and empirically tested, e.g. also in the research study of Kennedy and Fiss (2009), who controlled for the organizational size to reflect the speed of adoption. In an extensive research effort, Child (1973) provided empirical evidence on size, expressed through the number of employees, being a suitable predictor for organizational complexity and demonstrated in his analysis "strong links between size and complexity even when other contextual variables were taken into account" (Child, 1973, p. 181), leading to the conclusion that larger companies are subject to more bureaucratic processes and less flexibility. In a large meta-analysis of 80 empirical studies, Kimberly (1976) also finds that over 80 percent of these studies have used the number of employees to operationalize the internal structures of a company. Yet, with this pragmatic translation into number of employees, there is a risk of neglecting potential other implications of organizational complexity which might be expressed better through other measures, e.g. available resources or organizational output (Kimberly, 1976). To counter potential caveats, refraining from a binary perspective of whether size matters or not is recommended, and instead analysing "under what conditions aspects of size are determinants of dimensions of structure" (Kimberly, 1976, p. 594). Following the previous research efforts, the item of organizational size was included in the questionnaire. Main considerations with respect to the selected scale included the broad range of company sizes and the worldwide reach of the survey. The scale was intended to reflect the diversity of company sizes on a more granular level and the survey addressed an international target group. Thus, the enterprise classification scheme of the EU[36], distinguishing four size groups, seemed not adequately suitable for the

---

35   Author's illustration based on Hayes (2013).
36   The EU differentiates between provides a definition for SMEs based on three groups (micro, small, medium-sized) and also involves financial indicators to define the company size (https://ec.europa.eu/growth/smes/sme-definition_en).

present context. To capture size on a more detailed level, a scale was developed that consisted of eight potential groups[37] and included in the analysis.

Based on empirical evidence on the relationship of size and innovation implementation rather than innovation initiation (Damanpour, 1992), this measure was included as a moderating variable to test interaction effects with organizational factors. Following the suggestion to not merely include size as a binary variable, the ordinal scale allows for the variable to take a broad range of values and thus can be included in the following analysis as if it was continuous (Berry, 1993).

Covariates/ Control variables

Other than the variables listed above, the analysis also included control variables (or covariates), which help to more clearly isolate the effects of the independent variables on the dependent variable. They also contribute to the reduction of the variance which is not explained by the analysis (Stevens, 2002). As this group of variables helps to increase the interpretation accuracy, covariates are an important addition to the analysis and often cover aspects like age, gender or other demographic measures. Since the present analysis focused on the perceived success of C2C implementation, the model included four covariates reflecting potential influencing factors on a company-level, not concerning the respondent profile. The *current certification status* of the company (at the time of the survey participation) was measured on a binary Yes/No scale. The *age of the company*, expressed through the year of the company foundation, was included with an ordinal scale. The third covariate addressed the *most frequent C2C certification level* serving as an indication for potential implementation effort (Basic, Bronze, Silver, Gold, Platinum).

Building on proposition P6 (chapter 3.3), a construct reflecting the market success of the C2C certified product(s) was also included as a control variable. This construct, labelled *new product success,* might impact the relationship of the independent and dependent variable as it is related to the idea of satisfaction. Even though there are multitudinous studies on the success and failure of product development, such as the large-scale review by Cooper and Kleinschmidt (1987) or the meta-analysis by Evanschitzky et al. (2012), the operationalization of the concept 'new product success' proved to be challenging. The work of Lynn et al. (1999) provided an empirically tested construct consisting of eight items, out of which five were selected based on their fit to the present C2C context, and accordingly adapted in their wording. The respondents were asked to indicate their agreement on a 5-point Likert scale, ranging from '1 = Strongly disagree' to '5 = Strongly agree' (Table 8).

---

37 See chapter 4.3.1 for the detailed scale and the respective results.

Table 8: Survey items on organizational enablers – new product success

Item 'new product success'		Source
After the implementation of C2C standards, the company's C2C certified product(s)…		Lynn et al. (1999)
NPS1	Overall, met or exceeded sales expectations	
NPS2	Met or exceeded profit expectations	
NPS3	Met or exceeded return on investment (ROI) expectations	
NPS4	Met or exceeded market share expectations	
NPS5	Met or exceeded customer expectations	

4.1.2 Development of an online survey

Since the first part of the survey intended to set an apposite basis for the remaining questions on C2C implementation experience, the initial questions asked for concrete engagement measures of the company in the C2C certification program (Fink, 2013; Peterson, 2000). For example, the respondents were asked to indicate which year the first C2C certificate was received, which share of products were C2C certified or what the highest certification level of most of the products was. To reflect the implementation experience of both company types, the ones having abandoned but also the ones who continued the certification program, the survey was set up with two separate paths of questions. Hence, after the first question 'To your knowledge, does your company currently hold a C2C certificate for one or more products from your product portfolio?' an additional set of questions was integrated. The companies who answered with 'No' were asked when the last certification expired and which reasons led to the certification abandonment. Further, the survey items concerning the certified products were formulated in the past tense and the question whether the company planned to prolong or extend the certification were left out for this respondent group (Fink, 2013). All other questions, which addressed the certification and implementation experience, were kept the same for all companies. Finally, the respondents were asked about their personal background, e.g. functional department or tenure in the company, and the company background, e.g. size, industry or country of origin (Peterson, 2000)[38].

---

38   See Appendix 8.1 for the complete presentation of survey questions.

After the survey finalization, the next step comprised of the identification of C2C certified companies using the product registry of the C2C products innovation institute (www.c2ccertified.org). All listed companies were extracted at different points in time, in 2013, 2014 and 2015. As the analysis was also aiming to include companies which have left the certification program over time, an internet archive tool[39] facilitated the retrospective identification of companies which were holding the certificate before 2013 as well. This approach yielded a list of 205 (in the present or past) certified companies. Applying a nonprobability sampling method, a search for direct C2C contacts in the respective company was conducted to ensure that the survey questions could be answered by a C2C knowledgeable company representative (Bernard, 2013). As has already been revealed in the qualitative study, the numerous intersections of the C2C concept with different departments such as R&D, supply chain management or innovation management can result in different functional backgrounds of the C2C contact in charge depending on the company. To identify the relevant points of contacts, online search engines, company (sustainability) reports, press releases and professional online networks[40] were screened and analysed, resulting in 115 direct contacts and their contact details. To be able to address the companies without a clear C2C point of contact, a general contact form was used to ask for the person in charge of C2C.

This approach of using one single informant to reflect a company's standpoint on the subject matter is commonly used in business research. Key informants are a suitable data source for inter-organizational studies since they have both the expertise for the relevant subject matter and the willingness to openly share it. However, this procedure is also linked to a potential biasing of the responses, e.g. due to different management positions or subjective recalling of the queried events (Bryman and Bell, 2015; Kumar et al., 1993), which will be addressed in the course of bias treatment in chapter 4.5.2.

After the revision of all items in order to ensure that no jargon or technically complex language was used, phrases were unambiguous, and the order of the questions randomized, the survey was pre-tested for clearness and comprehensiveness with a group of academic colleagues as well as selected practitioners (Bernard, 2013). This resulted in minor wording adjustments and a slight shortening of the survey by excluding less critical items. These measures were undertaken to increase the responsiveness and decrease potential bias issues, such as common method bias (Bryman and Bell, 2015; Podsakoff et al., 2003). Furthermore, the pre-test was important in order to verify the technical viability of the online tool (Kuckartz et al., 2009). Even

---

39 https://web.archive.org
40 The used online networks were XING (www.xing.de), with a particular focus on German-speaking companies and LinkedIn (www.linkedin.com) for international companies.

though the single items were randomized in their order, the overall structure of the survey followed the 'funnelling' procedure, which traditionally moves from more general to more specific questions (Peterson, 2000). In the present study context, this approach provides a logical structure (see Figure 11) for the research focus by assisting the respondent in moving chronologically from the motivation to initiate C2C implementation, followed by the certification and implementation experience and closing by a concluding assessment (Kuckartz et al., 2009; Peterson, 2000).

Certification status
- Data collection on certification facts
- Two possible paths: currently certified 'yes' or 'no'

Motivational factors
- Specification of motives based on qualitative study and extant theory
- Free text entry to capture not considered company-specific factors

Implementation and certification process
- Internal and external enablers and barriers for C2C
- Assessment of the certification process

Outcome and concluding assessment
- Customer perception and intent to proceed with C2C
- Overall satisfaction

Company and respondent background
- Data collection on certification facts
- Classification data on company facts (at end of survey)

Figure 11: Structure of the online survey[41]

To be sent out to German, Austrian and Swiss companies, the questions were translated from English to German. In order to maximize the concordance between the two languages, the questions were first translated from English to German, then back into English by a native English speaking research team member. This way, the consistency and validity of both survey versions as well as comparability of results was ensured (Fink, 2013). With the help of the online tool SurveyMonkey[42], data collec-

---

41 Author's illustration.
42 www.surveymonkey.de

tion started in July 2015 when all companies were contacted at the same time, followed-up with up to two reminding e-mails until September 2015.

The survey link was sent out via e-mail. Using the customization functionalities of SurveyMonkey, it was possible to include a personal salutation of the participant in the introductory e-mail, e. g. "Dear Mrs. Smith", as well as the specific company name. This option was used in order to increase response reactions (Peterson, 2000). When the contact name had not been identified, the e-mail was sent out with the introduction "Dear Sir/Madam". In addition to a short description of the project and the actual survey link, the e-mail contained a personal introduction of the researcher, an indication of the necessary time to complete the survey and a confirmation of the anonymity of the survey responses (Figure 12).

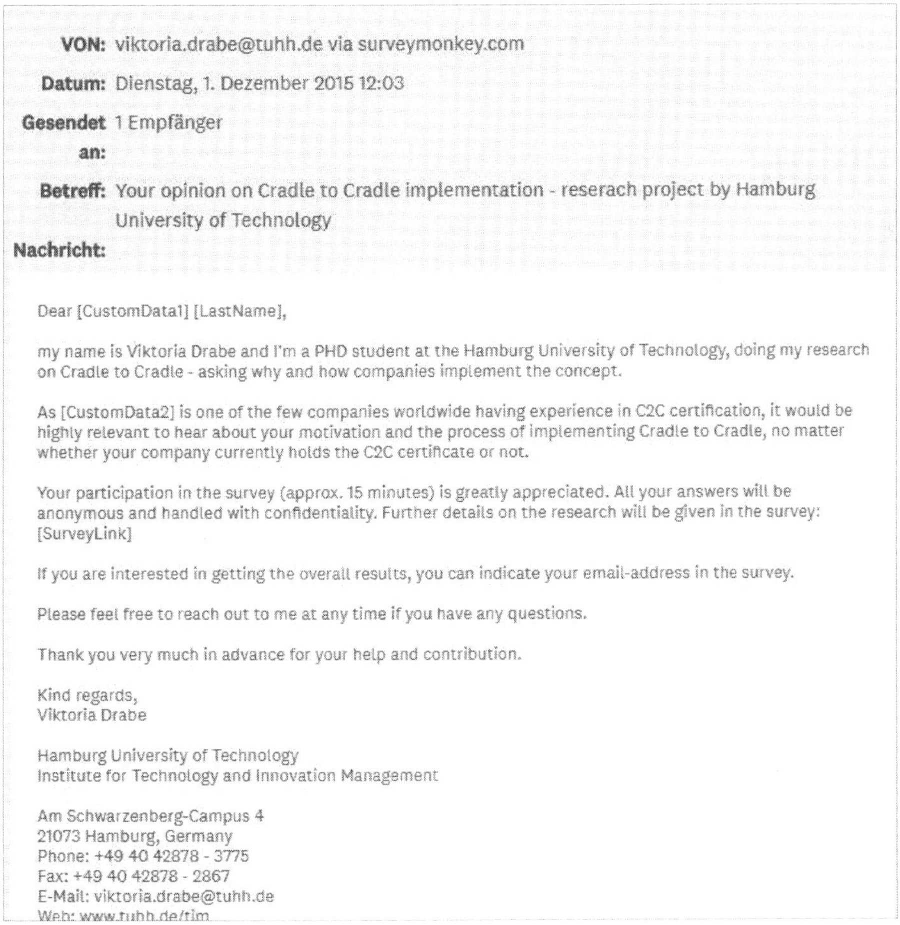

VON: viktoria.drabe@tuhh.de via surveymonkey.com
Datum: Dienstag, 1. Dezember 2015 12:03
Gesendet an: 1 Empfänger
Betreff: Your opinion on Cradle to Cradle implementation - reserach project by Hamburg University of Technology
Nachricht:

Dear [CustomData1] [LastName],

my name is Viktoria Drabe and I'm a PHD student at the Hamburg University of Technology, doing my research on Cradle to Cradle - asking why and how companies implement the concept.

As [CustomData2] is one of the few companies worldwide having experience in C2C certification, it would be highly relevant to hear about your motivation and the process of implementing Cradle to Cradle, no matter whether your company currently holds the C2C certificate or not.

Your participation in the survey (approx. 15 minutes) is greatly appreciated. All your answers will be anonymous and handled with confidentiality. Further details on the research will be given in the survey: [SurveyLink]

If you are interested in getting the overall results, you can indicate your email-address in the survey.

Please feel free to reach out to me at any time if you have any questions.

Thank you very much in advance for your help and contribution.

Kind regards,
Viktoria Drabe

Hamburg University of Technology
Institute for Technology and Innovation Management

Am Schwarzenberg-Campus 4
21073 Hamburg, Germany
Phone: +49 40 42878 - 3775
Fax: +49 40 42878 - 2867
E-Mail: viktoria.drabe@tuhh.de
Web: www.tuhh.de/tim

Figure 12: E-mail invitation for survey participation

After clicking on the survey link, the participant was introduced to the research background and was reassured about the confidential handling and processing of the provided data (Peterson, 2000). Finally, all respondents were offered the chance of receiving a consolidation of the anonymized survey results if they indicate their interest at the end of the survey (see Figure 13). Following the introductory page, the survey questions were displayed according to the aforementioned structure. After the concluding part on the respondent's personal and company background, the survey closed with the question whether the respondent would be interested in receiving the results or in participating in follow-up research projects concerning C2C (Kuckartz et al., 2009).

Figure 13: Title page of online survey

With respect to concrete formulation and the corresponding scales, the construction of the questionnaire followed established guidelines in order to ensure the obtained data was reliable and valid. Besides the common considerations regarding a simple, brief and unambiguous wording, this also implies a preceding analysis of the targeted respondents and their expected level of knowledge, e. g. of specific terms. Further, it is critical to apply neutral and objective wording to avoid biases in answers (Iarossi, 2006; Peterson, 2000; Podsakoff et al., 2003). The rating scales were developed based on the single questions, in most of the cases using the widely established Likert scale with five levels of scores. No scales were introduced with numerical values only but were always complemented with a clear verbal label in order to exclude ambiguity or misconception (Tourangeau et al., 2000). Where applicable, a specific scale was adopted from previous research studies or precisely developed for the C2C context. The latter was the case for the introductory questions on the company's status of C2C engagement, e. g. level of certifications or share of certified products. In addition to the applied scales, the option 'No answer' was available for every question (Sekaran and Bougie, 2010).

4.2 Data preparation for statistical analysis

4.2.1 Data review and cleansing

Once the data collection was finalized, the survey link was closed and the data exported from the survey tool into an Excel format for further revision. Of the 205 addressed C2C companies, 22 German and 57 English questionnaires were filled out, leading to a total of 79 returned surveys. To be able to perform further analyses, the German and English response sets had to be merged and the full set of responses double-checked. To facilitate comparability of results and ensure a uniform basis for analysis, all German answers were translated into English ones, which presented no issue in most cases since the scales were all assigned to numbers, such as '1 = Strongly disagree'. In the last part, when answers were assigned to numbers based on their alphabetical order, e. g. the country of origin ('1 = Belgium' in German language while '1 = Austria' in the English questionnaire) or the industry sector, the German answers had to be adequately transferred into the English equivalent. Furthermore, the reversely coded items were adequately adapted. In a final step, the uncompleted surveys were identified and removed from the sample, leading to a final set of 72 responses.

4.2.2 Evaluation of missing Data

As in the first cleansing step, incompleteness was not judged based on a few single missing answers but an obvious early termination of the survey, mostly after the first few questions, the second step comprised of a thorough analysis of missing values to omit potential subsequent biases. Following the four-step approach of Hair et al. (2014), the type and extent of missing data were determined leading to the exclusion of all variables with missing data in more than 10 % of the cases. Thus, in total seven variables were excluded due to missing values ranging between 11.1 % and 15.3 %. Then, the randomness of the missing data in the remaining variables was analysed in order to be able to select an appropriate imputation method. For this purpose, the Little's MCAR (Missing Completely At Random) test was conducted which tests whether "the missingness does not depend on the data values" (Little and Rubin, 2002, p. 12). With a significance value of 0.747 (Chi-Square = 1341,672, DF = 1377), the test, performed with IBM SPSS Statistics 23, was not significant meaning that the data is missing completely at random and missing values can be imputed with a suitable method (Hair et al., 2014; Little and Rubin, 2002). Building on Hair et al. (2014), in a following step, an imputation method had to be selected so that the missing values for the remaining variables (with missing data below 10 %) could be replaced. A suitable and established approach, also specifically for exploratory research, is the Expectation Maximization (EM) algorithm. For missing data to be imputed with this method, the statistical non-significance of the MCAR test has to be confirmed first (Hair et al., 2014), which has been performed in the previous step. In the EM algorithm a series of steps is performed iteratively in order to achieve convergence: "(1) replace missing values by estimated values, (2) estimate parameters, (3) re-estimate the missing values assuming the new parameter estimates are correct, (4) re-estimate parameters, and so forth" (Little and Rubin, 2002, p. 166). The imputation was performed using IBM SPSS Statistics 23, eventually leading to the replacement of missing values for the items used in the successive factor analysis. No values were replaced for questions from the first survey part on the company's certification details and the last part on the respondent and company profile, either no values have been replaced for the items related to the dependent variable.

4.3 Descriptive analysis

In the following section, a descriptive analysis of the survey data provides an overview of the data set by analysing respondent and company profiles. As mentioned in the previous section, from the 205 companies worldwide that received the online questionnaire, 79 surveys were returned. After extracting the incomplete surveys, a total of 72 usable surveys remained in the dataset yielding a response rate[43] of 35 % (see Figure 14). Out of the final sample 64 companies (89 %) were holding a C2C certificate at the time of the survey whereas eight responding companies (11 %) were not certified any more when they filled out the survey.

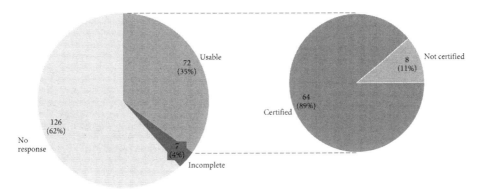

Figure 14: Overview of survey respondent set[44]

4.3.1 Company profiles

As explained earlier (chapter 2.3.3), the C2C certification program consists of a progressive logic with certification levels ranging from Basic to Platinum. Since the certificate is granted on a product basis, this also implies that companies can have different products with different certification levels. To get a better understanding of the company's experience with certification, companies were asked for the C2C certification level that "most of the company's certified products have" (Table 9).

---

[43] According to a large-scale meta-analysis of Baruch and Holtom (2008), the average response rate of surveys targeting organization is 35.7 % with a standard deviation of 18.8.
[44] Author's illustration.

Table 9: C2C certification level of most of the company's certified products

Certification level	Frequency	Percent
Basic	8	11,1
Bronze	14	19,4
Silver	44	61,1
Gold	5	6,9
Platinum	0	0,0
N/A	1	1,4
Total	72	100,0

For 61% of the companies, the most frequent level was silver, followed by the bronze (19%), basic (11%) and gold level (7%). No platinum certification[45] was present in the sample.

Concerning the company experience with C2C, the responses reveal a rather diverse picture. Asked for the share of C2C certified products in the total number of all products, two extremes could be observed. On the one hand, almost 40% stated to have less than 10% C2C certified products within the company's portfolio. On the other hand, in 25% of the responding companies, the C2C certified products accounted for more than 70% of all products.

Reflecting the various types of companies that were engaging in the C2C certification program, the sample was intended to uncover a granular picture of C2C certified companies with respect to their size and age (Table 10).

---

45 At the time the survey was conducted, there was no listed platinum certification. According to current status of the C2C product registry, one company holds a platinum certification (Rajby Textiles PVT Ltd) (see https://www.c2ccertified.org/products/registry, accessed 15.11.2020).

Table 10: Size and year of foundation of respondent companies

		Frequency	Percent
Company size	1–10 employees	2	2,8
	11–50 employees	11	15,3
	51–100 employees	9	12,5
	101–500 employees	25	34,7
	501–1.000 employees	5	6,9
	1.001–5.000 employees	9	12,5
	5.001–10.000 employees	3	4,2
	More than 10.000 employees	6	8,3
	N/A	2	2,8
Year of company foundation	1900 or earlier	16	22,2
	1900–1950	19	26,4
	1951–1980	14	19,4
	1981–2000	11	15,3
	2001–2010	9	12,5
	2011 or later	1	1,4
	N/A	2	2,8
	Total	72	100,0

The sample included micro and small companies as well as mid-sized and also large companies. Concerning the company age, there was a clear dominance of established companies with their company foundation dating back at least 30 years ago (68 %). These results indicate that the C2C certification is not a phenomenon that more strongly attracts start-up firms but presents a relevant concept for established and mid-sized companies. Looking at the point in time at which the companies started C2C certification, it also becomes apparent that the early adopters of the C2C concept were often the older companies. At the beginning of the certification program (before and in 2008) 82 % of the companies who adopted the certificate were founded before 1980. In general the development of the certification shows a decline after 2013, so that most of the companies (88 %) in the sample have been familiar with the concept for more than two years (Figure 15).

Quantitative exploration of motivational factors and organizational enablers

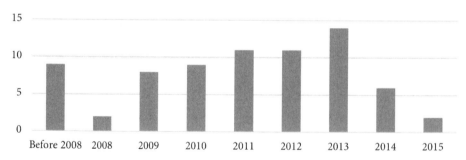

Figure 15: Year in which the company received the first C2C certificate[46]

With respect to the industry, the responding companies operate in various sectors. The building materials as well as the textile and fabrics sectors are the most represented ones in the sample (16 and 15 companies). The number of companies with a B2B (business-to-business) commerce focus is more than double the number of companies operating with end consumers (Table 11).

Table 11: Industry and commerce focus of respondent companies

Industry in which the company operates*	Frequency	Commerce focus*	Frequency
Building materials	16	Business-to-business (B2B)	49
Textile and fabric	15	Business-to-customer (B2C)	22
Paper & Print	8	Business-to-government (B2G)	2
Furniture	7	N/A	2
Chemicals	5		
Floor covering	5		
Interior design	5		
Packaging	5		
FMCG	4		
Office supplies	2		
Personal Care	2		
Lighting	2		
Home Care	1		
Other	9		
*Multiple responses possible			

---

46   Author's illustration.

The geographical focus of the C2C diffusion appears in the Netherlands (29 %) and the USA (24 %), accounting in sum for more than half of all companies in the sample. 11 companies (15 %) came from Germany (Table 12).

Table 12: Country of origin of respondent companies

The company's country of origin	Frequency	Percent
Netherlands	21	29,2
USA	17	23,6
Germany	11	15,3
Belgium	5	6,9
Denmark	4	5,6
Switzerland	4	5,6
Taiwan	2	2,8
Austria	1	1,4
Spain	1	1,4
United Kingdom	1	1,4
Israel	1	1,4
N/A	4	5,6
Total	72	100,00

Comparing this dispersion with the initial set of 205 companies which were listed in the C2C certified products registry and which received the survey, the same three countries are dominant, however with the difference that the largest number of companies comes from the USA, not from the Netherlands. This small difference between the overall and the respondent sample might be due to the closer proximity of the Netherlands to Germany, which might result in a greater openness towards the survey. Still, the set indicates a suitable representation of the geographical C2C diffusion.

The survey also asked for a company's future intentions regarding the C2C certification. From the currently certified companies, 72 % indicated that they would like to prolong the current certification, 26 % were not sure and 2 % stated their decision to quit the certification program. In order to gain a better understanding of whether the C2C concept can be applied to a broader product range within a company, the respondents were also asked to indicate a potential extension of the certification. Here, less than half of the companies (44 %) responded positively and almost the same number of respondents (37 %) were undecided. The rest of the companies (19 %) did not plan an extension of the certification to other products (Figure 16).

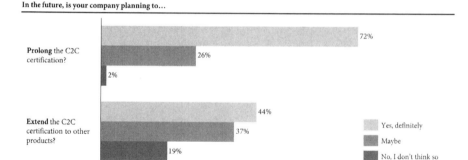

Figure 16: Future plans to prolong or extend C2C certification[47]

In contrast to the inquiry for future intents of currently certified companies, the respondents which had stated at the beginning that they are not certified any more, were asked for the reasons for abandoning the C2C certification program. Besides some pre-formulated response options, the respondents could also use comment fields to include individual answers. For the major part of the companies (44%), the cost of the certification was the dominant motive for quitting the certification. A more detailed look into the company specifics, such as company size or age, did not show any further pattern, so that this argument could not be linked to a specific company type. In the comment section, the respondents articulated additional details, which were mostly related to the certification effort, such as "demand for extra data for re-certification even when nothing has changed with the product is too high" as well as a lack of positive customer response, e. g. "customers do not care at all" (Figure 17).

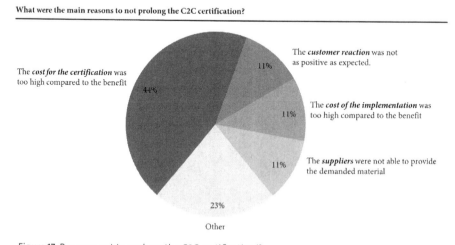

Figure 17: Reasons not to prolong the C2C certification[48]

47 Author's illustration.
48 Author's illustration.

4.3.2 Respondent profiles

Of the total sample of respondents, the majority was male (71%), 17 respondents were female (24%) and 5% did not indicate their gender. The age was distributed normally (see Figure 18), with 41–50 years being the largest group (32%) and indicating a certain seniority level of the respondents.

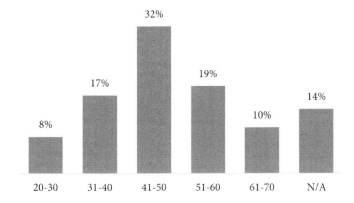

Figure 18: Age distribution of respondents (years)[49]

In order to better assess the knowledgeability of the respondents concerning internal processes and the C2C implementation, the respondent's tenure with the company as well as in the current position were included in the survey. 51% of the respondents indicated having been with the company for over 11 years. Another 13% had belonged to their company for 8–11 years. The remaining respondents had worked in their companies for 3 years or less (17%), and 4–7 years (17%). Therefore, the numbers can be an indication for the familiarity of the respondents with organizational processes and routines. The average tenure in number of years is lower for the respondents' current position, in which the majority (26%) had worked for 4–7 years. Table 13 summarizes the distribution of the respondents' tenures.

---

49 Author's illustration.

Table 13: Tenure of respondents

	0–3 years	4–7 years	8–11 years	More than 11 years	N/A	Total
With the company						
number of respondents	12	12	9	37	2	72
in percent	16,7	16,7	12,5	51,4	2,8	100
In current position						
number of respondents	16	19	16	16	5	72
in percent	22,2	26,4	22,2	22,2	6,9	100

The respondents' relatively high average number of years within the company is also reflected in their professional experience. In total 75 % of the respondent set belonged to the middle (35 %) or top (40 %) management in their company. This suggests a certain decision power, which is particularly relevant when looking at company motivations for C2C engagement. Respectively 11 % indicated having a position in the lower management or as an employee, 3 % did not indicate their position level.

The survey participants were selected and addressed based on their experience with the company's C2C certification, not on a specific functional department. As illustrated in Table 14, the three largest groups were respondents belonging to the Sustainability department (24 %), the executive board (18 %) and the R&D department (18 %).

Table 14: Respondents' functional background

Functional department in the company	Number of respondents	Percent
Sustainability	17	23,6
Executive Board	13	18,1
Research & Development	13	18,1
Marketing	8	11,1
Sales	7	9,7
Innovation Management	3	4,2
Production	3	4,2
PR / Communication	1	1,4
Other*	5	6,9
N/A	2	2,8
Total	72	100

*Specified as: Combined Sales, R&D, Sustainability (1); Manager Energy & Environment (1); Product Management (2); Quality & Standards (1)

Contributing to the trans-disciplinary character of C2C, the selection process of the survey recipients (as elaborated in chapter 4.1.2) was not limited to functional departments. The diverse picture of the respondents' functions confirms this trait and further suggests a certain strategic relevance of C2C as the second largest group of respondents belonged to the Executive Board.

4.3.3 Descriptive analysis of main variables

The study intended to shed more light on the motivations of companies to engage with a sustainability-related practice, precisely the C2C certification program, and their implementation experience. Therefore, the survey participants were asked to rate the importance of the motivational factors related to their C2C engagement. Figure 19 illustrates the levels of agreement, distinguished according to the responding company's current certification status (whether they were C2C certified at the time of the survey or not) and the average values per item. The motive with the highest average importance level (mean = 4.28) was the fit of C2C with the corporate philosophy. At the same time, the values reflect that for the companies which abandoned the certification, this motive was of lesser importance. The motivation with the second highest average score (mean = 4.25) was the 'expectation to be perceived as a market leader'. In this case the importance of this motive was ranked even higher by the companies which didn't extend their certification than by the ones who did. This might suggest a pioneering prospect that might have remained unfulfilled. The third highest mean (mean = 4.11) corresponds again to a factor which expressed an expectancy of benefits, precisely 'the expectation of C2C to be a source of new opportunities'.

With respect to the lower importance ratings, the items related to the motives of legitimation or threat avoidance stand out and suggest that the initial motivation of the surveyed companies was rather linked to potential benefits and business opportunities than being a reactive approach to the company environment and legitimation. More in general, the answers also reveal a wide spread in the individual importance ratings of motivations between the companies.

Descriptive analysis

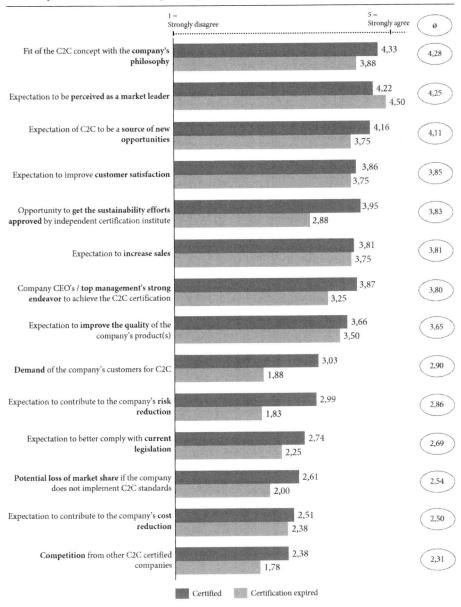

Figure 19: Averages for motives for C2C implementation and certification[50]

---

50 Author's illustration.

When asked for their conclusive assessment of the product success, the spread of answers was not as wide, ranging from mean values between 2.92 and 3.60, with the unfulfilled expectations regarding the return on investment being the variable with the lowest agreement score (see Figure 20). The customer expectations were agreed to have been met or exceeded at a higher level, interestingly with only minor differences between certified and non-certified companies. A higher difference in agreement for met or exceeded expectation was related to profit and sales of the certified products. Here the non-certified companies had indicated their agreement with a difference of 0.68 compared to the certified ones.

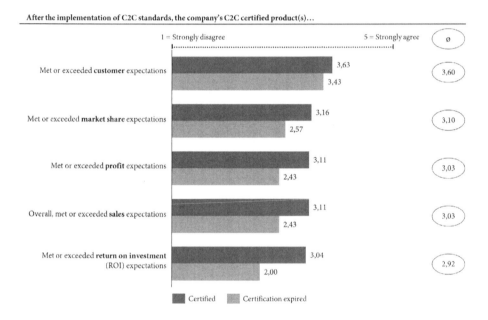

Figure 20: Review of met or unmet expectations[51]

After reviewing the final evaluation with respect to the initial expectations and in how far they have been met, the companies were also asked for their agreements with a more general, conclusive statement on their experience with C2C (see Figure 21). All of the statements have an average rating higher than 3.50 (based on the Likert scale from 1 to 5) already suggesting a certain degree of satisfaction, which needs to be further analysed in the course of the regression analysis. When looking at the differences between the ratings of certified and non-certified companies, two variables show particularly interesting results. First and probably not surprisingly, for the vari-

---

51   Author's illustration.

able 'we would recommend the implementation of C2C to our business partners', the companies having quit the certification program have an agreement value of 2.86 compared to still certified companies with a mean value of 3.60, which presents the highest difference between certified and non-certified companies[52].

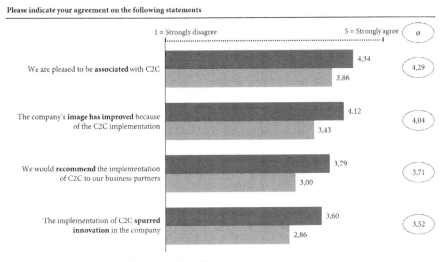

Figure 21: Final evaluation of C2C experience[53]

Second, the reversely coded variables addressing the cost of certification and implementation, indicate a certain discontent with the cost-benefit trade-off (see Figure 22). For the average agreement rating with respect to the cost of C2C implementation, the assessment is more negative for respondents with an expired certification (mean = 3.57) than for the ones currently certified (mean = 3.50).

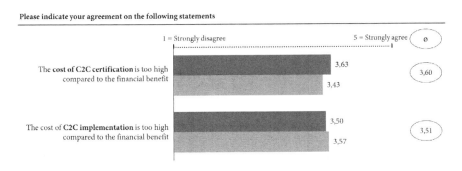

Figure 22: Assessment of certification costs[54]

---

52 Further descriptive results on the responding companies are illustrated in Appendix 8.2.
53 Author's illustration.
54 Author's illustration.

4.4 Exploratory Factor Analysis

Given the lack of empirical evidence in the context of Cradle to Cradle, an exploratory factor analysis (EFA) was the method of choice in order to explore the latent factors influencing the dependent variable 'satisfaction with C2C implementation'. Especially in the early phases of evolving research streams, the EFA, in contrast to the confirmatory factor analysis (CFA), presents an appropriate method to test and further develop existing scales by adding new scales (Hurley et al., 1997; Wolff and Bacher, 2010). As the survey also contained variables which resulted from the qualitative study, the EFA was selected to be highly suitable to cast more light on the underlying structures among the variables and identify the most influential factors (Weiber and Mühlhaus, 2014). The results of the analysis then provide the basis for the subsequent regression analysis[55].

4.4.1 Testing underlying assumptions

Before conducting the exploratory factor analysis, the key underlying assumptions need to be checked to ensure the suitability of the approach. In doing so, it is recommended to not only focus on the statistical requirements, which will be explained in the following, but also to ensure that the analysis is conceptually relevant (Hair et al., 2014). All following statistical analyses were performed using the IBM SPSS Statistics 23 program.

Conceptual relevance

Prior to performing the statistical analyses, the data set should be thoroughly scrutinized to ensure the conceptual suitability of the explorative factor analysis in the present study context (Hair et al., 2014; Hurley et al., 1997). Hereby, one critical factor is the assumption that an 'underlying structure' does actually exist among the included variables (Hair et al., 2014) as the analysis intends to identify these potential structures and to systemize them by clustering the variables according to their correlations (Wolff and Bacher, 2010). There are various reasons which underline the suitability of the present data set for EFA. On the one hand, it is based on the combination of theoretically more established factors as well as new variables resulting from the qualitative exploration in the first study (chapter 3), thus it is not founded on comprehensive existing empirical evidence. On the other hand, the qualitative

---

55 In the following sections, the variable names are shortened or simplified to improve legibility and clarity. The verbatim formulation of the respective items from the online survey can be found in chapter 4.1.1.

study already indicated some potential patterns with respect to influencing factors that shall be further examined with the help of the EFA. Since no concrete structures or specific numbers of factors have been detected yet, the factor analysis provides an appropriate means of addressing this deficit (Wolff and Bacher, 2010). To be able to more profoundly explore the critical influencing factors, the variables were split in two groups, the motivational and organizational context variables, and analysed separately in the EFA. As described earlier, both groups of variables consisted of established constructs as well as newly developed variables from the qualitative study. The items intended to operationalize the dependent variable (company satisfaction with C2C implementation) were treated as a separate set and third group in the EFA.

Sample size for EFA

The academic discussion about appropriate sample sizes for quantitative analyses are controversial and discordant concerning necessary thresholds and requirements (e.g. Williams et al., 2010; Conway and Huffcutt, 2003). In their reviews, the authors summarize different approaches towards the assessment of an appropriate sample size, ranging from a sufficient number of 50 up to a required minimum sample of up to 1000 or greater (Williams et al., 2010). Due to this variation in possible sample size requirements, the present analysis follows the recommendation of MacCallum and Widaman (1999) to also take into account the various application possibilities of factor analysis and not to use generic rules of thumb. When considering the study context for the identification of an appropriate threshold, the respective circumstances may also allow relatively small sample sizes, e.g. when communalities are sufficiently high (> 0.6) or if factors are not based on single items, which both applies for the present study. Further, the minimum number of 50 observations is also recommended as the lower bound of an EFA by Hair et al. (2014). They specify this minimum with the requirement of having "more observations than variables" (Hair et al., 2014, p. 100). Both conditions are met with the present sample size of 72 responses.

Intercorrelation measures

One critical indication for the presence of an underlying variable structure can be found by means of different correlation analyses. In the case of no correlations between the variables, an exploratory factor analysis would be rendered nugatory as no factors could be produced based on the single items (Backhaus et al., 2016; Hair et al., 2014). In addition to a visual inspection of the correlation tables (Table 15 and Table 16), which indicated numerous significant correlations, three established analytical tests were performed to ensure a sufficient level of correlation: the *Bartlett test of sphericity,* the *measure of sampling adequacy (MSA)* and the *Kaiser-Meyer-Olkin* (KMO) criterion.

The Bartlett test of sphericity "provides the statistical significance that the correlation matrix has significant correlations among at least some of the variables" (Hair et al., 2014, p. 102), hence the test determines whether the correlation matrix shows significant correlations and the variables actually are suitable for an exploratory factor analysis. If the correlation coefficients are significantly (< 0.05) higher or lower than 0, the intercorrelations can be appropriately analysed further in the EFA (Dziuban and Shirkey, 1974).

Introducing an index, which ranges from 0 to 1 (with one being an accurate prediction of one variable by the others) the measure of sampling adequacy (MSA) also quantifies the suitability of the variable intercorrelations for further EFA, which is presented by one overall MSA value outcome as well as variable-specific values in the anti-image correlation matrix for every included item (Hair et al., 2014). The commonly applied guideline for this measure classifies different qualities of the variable set: 0.80 or above – meritorious, 0.70 or above – middling, 0.60 or above – mediocre, 0.50 or above – miserable, below 0.50 – unacceptable (Hair et al., 2014; Kaiser and Rice, 1974; Weiber and Mühlhaus, 2014). In order to increase the overall MSA value, Hair et al. (2014) recommend excluding all variables with MSA values < 0.50 from the factor analysis.

Building on an aggregation of the MSA values, the Kaiser-Meyer-Olkin measure evaluates the accuracy of the extracted structures of relationships and is also evaluated based on a threshold ranging from 0 to 1 (Kaiser and Rice, 1974). The underlying thresholds correspond to the ones from the MSA analysis based on the work of Kaiser and Rice (1974), meaning that the higher the KMO value, the more adequate the data sample for an EFA. A more conservative approach towards the exclusion of variables is provided by Dziuban and Shirkey (1974) who recommend a minimum KMO value of 0.6. The results for the presented intercorrelation measures are elaborated in the following two chapters.

Exploratory Factor Analysis

Table 15: Descriptives and correlations for the group of motivational variables

Variables		Mean	Std. Deviation	M1	M2	M3	M4	M5	M6	M7	M8	M9	M10	M11	M12	M13	M14
M1	Fit of C2C with the company's philosophy	4.28	0.890	1													
M2	Opportunity to get the company's sustainability efforts approved by an independent certification institute	3.83	1.077	.253*	1												
M3	Company CEO's/top management's strong endeavor to achieve C2C certification	3.80	1.303	.528**	.297*	1											
M4	Expectation to contribute to the company's cost reduction	2.50	1.186	.341**	.243*	.358**	1										
M5	Expectation to contribute to the company's risk reduction	2.86	1.167	.354**	.431**	.421**	.508**	1									
M6	Expectation to increase sales	3.81	0.929	.135	.136	.304**	.408**	.448**	1								
M7	Expectation of C2C to be a source of new opportunities	4.11	0.972	.431**	.251*	.325**	.311**	.416**	.383**	1							
M8	Potential loss of market share if the company does not implement C2C	2.54	1.291	.128	.316**	.292*	.595**	.534**	.336**	.316**	1						
M9	Competition from other C2C certified companies	2.31	1.157	.120	.334**	.159	.533**	.432**	.288*	.192	.611**	1					
M10	Expectation to improve the quality of the company's product(s)	3.65	1.301	.465**	.159	.280*	.502**	.501**	.385**	.455**	.311**	.187	1				
M11	Expectation to be perceived as a market leader	4.25	0.944	.388**	.333**	.396**	.419**	.300*	.375**	.399**	.273*	.062	.416**	1			
M12	Expectation to improve customer satisfaction	3.85	0.989	.140	.311**	.220	.404**	.510**	.489**	.447**	.349**	.175	.367**	.389**	1		
M13	Demand of the company's customers for C2C	2.90	1.313	.033	.239*	.030	.183	.374**	.262*	.297*	.499**	.321**	.183	-.003	.336**	1	
M14	Expectation to better comply with current legislation	2.69	1.289	.346**	.361**	.211	.483**	.567**	.348**	.393**	.514**	.387**	.499**	.253*	.362**	.473**	1

* Correlation is significant at the 0.05 level (2-tailed).
** Correlation is significant at the 0.01 level (2-tailed).

Quantitative exploration of motivational factors and organizational enablers

Table 16: Descriptives and correlations for the group of organizational context variables

	Variables	Mean	Std. Deviation	IMP1	IMP2	IMP3	IMP4	IMP7	PAR1	PAR2	PAR3	PAR4	PAR5	PAR6	PAR7	TS4	TS5	TS6	TS7	CC2	CC4
IMP1	The company has integrated C2C in procedures and work instructions	3.59	1.012	1																	
IMP2	The company has identified specific persons and positions responsible for C2C implementation	4.28	0.632	.399**	1																
IMP3	The company keeps records of the training provided to staff in relation to the implementation of C2C	3.23	1.133	.509**	.361**	1															
IMP4	The company has obliged its supply base to supply according to C2C	3.49	1.211	.485**	.130	.497**	1														
IMP7	The extent to which you believe that at this point in time C2C philosophy, standards, and methods have been implemented throughout your company	3.81	0.999	.583**	.351**	.258*	.390**	1													
PAR1	We feel indebted to EPEA and/or MBDC for what they have done for us	3.17	1.019	.336**	.121	.215	.175	.319**	1												
PAR2	The company's employees share close social relations with the employees from EPEA and/or MBDC	2.90	1.073	.471**	.475**	.268*	.195	.373**	.510**	1											
PAR3	Our relationship with EPEA and/or MBDC can be defined as "mutually gratifying"	3.44	0.926	.350**	.244*	.345**	.242*	.294*	.626**	.443**	1										
PAR4	We expect that we will be working with EPEA and/or MBDC far into the future	3.89	0.962	.207	.025	.101	.392**	.312**	.456**	.345**	.615**	1									
PAR5	The collaboration with suppliers was extremely important to implement C2C	3.90	1.190	.345**	.093	.334**	.455**	.130	.224	.252*	.273*	.166	1								
PAR6	The company extensively collaborated with scientific partners, e.g. universities, in order to implement C2C	2.74	1.115	.335**	-.020	.242*	.165	.323**	.428**	.380**	.365**	.209	.324**	1							
PAR7	The company extensively exchanged knowledge with other companies in order to implement C2C	3.13	1.210	.215	.060	.153	.137	.131	.316**	.340**	.462**	.283*	.403**	.674**	1						
TS4	The company's engineering resources were more than adequate for implementing C2C	3.06	1.208	.566**	.206	.438**	.262*	.375**	.276*	.262*	.312**	-.007	.206	.404**	.233	1					
TS5	In the course of implementing C2C, the company made fundamental changes to the existing product and processes	2.97	1.189	.357**	.155	.425**	.417**	.240	.497**	.255*	.473**	.245*	.198	.347**	.199	.610**	1				
TS6	The innovation process had to be significantly adapted to C2C	3.19	1.273	.379**	.130	.325**	.624**	.342*	.301*	.267*	.434**	.475**	.300*	.363**	.312**	.358**	.530**	1			
TS7	The implementation of C2C had a significant impact on the entire value chain of the company	3.84	0.890	.217	.069	.004	.132	.180	.308**	.303**	.175	.072	-.091	.019	.093	.154	-.006	-.019	1		
CC2	The top management provided guidance during the implementation of C2C	3.75	1.054	.269*	.209	.258*	.478**	.292*	.208	.180	.314**	.350**	.117	.168	.236*	.319**	.311*	.403**	.190	1	
CC4	Did the company's internal stakeholder (e.g. marketing, product development, sales) show resistance to the implementation of C2C?	3.91	1.031	-.110	-.152	-.194	.066	-.050	-.137	-.076	.036	.107	-.253*	-.232*	.028	-.220	-.130	-.052	.097	.180	1

*. Correlation is significant at the 0.05 level (2-tailed).
**. Correlation is significant at the 0.01 level (2-tailed).

4.4.2 Extraction and factor rotation

Depending on the concrete goal of the EFA, there are different approaches to extract the final number of factors from the variable set, namely the principal component and the common factor analysis, as well as to rotate these factors for a more accurate interpretation.

Extraction method
There is no established consensus in research regarding which extraction method is generally better or more suitable. Instead, the application depends on the specific research context and data interpretation and sometimes even produces similar results (Henson, 2006). The primary objective being the reduction of data, aiming to explain a maximum portion of the total variance with a minimum number of factors, the principal component analysis (PCA) was selected as the most suitable extraction method (Hair et al., 2014) for the study at hand. This extraction method "considers the total variance and derives factors that contain small proportions of unique variance and, in some instances, error variance" (Hair et al., 2014, p. 105). In contrast to the common factor analysis, the principal component analysis includes the full variance into the factor matrix, not only the common variance (Hair et al., 2014). Therefore, based on the present study aim being data reduction and since the PCA offers a more accurate approximation towards the identification of underlying variable relationships with fewer principal components, it is selected as extraction method for the following factor analyses.

Rotation method
To facilitate the interpretation of the results, the choice of an appropriate rotation method is critical too. There are two main categories of rotation strategies, the *oblique* and the *orthogonal* rotations. The first method refers to an oblique rotation where factors can be correlated with one another and the latter describes an orthogonal rotation process assuming no correlations among the factors at all (Hair et al., 2014). Owing to the fact that a correlation between the factors is assumed in the present case, an oblique rotation method is the more suitable strategy. There are different methods of oblique rotation that the researcher can choose from, often producing comparable results[56]. One of the most widely used methods is the Promax rotation, which is also supported by the IBM SPSS Statistics 23 program and hence applied in the following analyses (Jennrich, 2006; Weiber and Mühlhaus, 2014).

---

56 See Jennrich (2006) for a detailed review of rotation methods.

4.4.3 Results of analysis

In addition to the aforementioned requirements related to the intercorrelations and the actual factor loadings, the *communalities* of the variables and the *eigenvalues* of the factors are critical for the final number of extracted factors. The communality indicates the extent of explained variable variance which the factor solution accounts for and ranges between 0 and 1 (Hair et al., 2014). As a common cut-off value, at least half of the variance of each variable should be taken into account by the factor solution, which means that variables with a communality lower than 0.5 should be excluded from the analysis (Hair et al., 2014; Weiber and Mühlhaus, 2014).

With respect to the significance of the factor loadings, there are various guidelines and approaches suggested by researchers, depending on the context and objective of the analysis (Stevens, 2002). For example, Hair et al. (2014) differentiate between practical and statistical significance, suggesting 0.5 as a minimum factor loading for a factor to be practically significant. The statistical significance also links factor loading to sample size, hence the required factor loadings decrease with a growing sample size. This recommendation is amplified by a consideration of the number of variables, which points at the necessity of accurately evaluating each individual factor solution depending on various criteria (Hair et al., 2014). While loadings can be statistically significant in larger samples, however, this would not necessarily mean that the variable and the factor have a high shared variance, which puts more focus on the consideration of practical significance (Stevens, 2002). At the same time, the evaluation of factor loadings should not only focus on the highest loading but also consider how much lower the other loadings are. In this context the indicated minimum loading value of 0.4 is regarded as moderate and sufficient (Gerbing and Anderson, 1988). In this regard, in order to ensure rigour of the analysis, the lack of a clear consensus on significant factor loadings is taken into account by an accurate assessment of several conditions, such as the loading values for the other factors, the cumulative variance explained and a critical factor loading value of at least 0.5.

For the following analyses, the results of the factor pattern matrix, presenting the unique contribution of each variable to the single factors, will be elaborated in detail. Yet, the correlations between variables and factors outlined in the factor structure matrix were also taken into consideration for the interpretation of the final results (Hair et al., 2014). In the course of the factor analysis only factors with eigenvalues greater than 1 were extracted since this value indicated whether the underlying factor provided a superior explanation of the variable's variance than the single variable would, hence this condition is regarded necessary for the factor significance (Hair et al., 2014). All variables were standardized before the calculation and all requirements were evaluated based on the criteria developed in chapter 4.4.1.

Motivational factors

The exploration of the items on the company's motivations resulted in three factors. After the initial assessment of all 14 motivational variables, the variable M2 was excluded due to an insufficiently high communality value (0.367). The remaining 13 motivators were included in the exploratory factor analysis, showing MSA values above the required threshold of 0.5, a 0.000 significance for the Bartlett test of sphericity and a KMO measure of 0.836 (Table 17).

Table 17: MSA and communalities for motivations – after deletion of M2

Item		MSA	Communalities
M1	Fit of C2C with the company's philosophy	.715	.682
M3	Company CEO's/top management's strong endeavor to achieve C2C certification	.763	.581
M4	Expectation to contribute to the company's cost reduction	.853	.651
M5	Expectation to contribute to the company's risk reduction	.898	.630
M6	Expectation to increase sales	.870	.516
M7	Expectation of C2C to be a source of new opportunities	.892	.548
M8	Potential loss of market share if company doesn't implement C2C	.807	.739
M9	Competition from other C2C certified companies	.783	.742
M10	Expectation to improve the quality of the company's product(s)	.875	.539
M11	Expectation to be perceived as a market leader	.840	.565
M12	Expectation to improve customer satisfaction	.836	.689
M13	Demand of the company's customers for C2C	.751	.612
M14	Expectation to better comply with current legislation	.901	.563
Kaiser-Meyer-Olkin Measure of Sampling Adequacy		.836	
Significance of Bartlett's Test of Sphericity		.000	

The analysis indicated a three-factor solution with all extracted factors showing significant eigenvalues greater than 1, a total variance explained exceeding the recommended value of 0.6 (Hair et al., 2014), however several factor loadings were lower than the critical threshold of 0.5. These indices call for a further exclusion of variables in order to increase the significance of the results. Table 18 displays the factor loadings for the three-factor solution and points at three variables, which lack significant loadings on any of the factors, greyed out in the table.

Table 18: Factor loadings of motivations – before deletion

	Pattern Matrix Factor		
Item	1	2	3
M1	.027	-.086	**.851**
M3	.122	-.082	**.754**
M4	**.606**	.000	.387
M5	.445	.342	.202
M6	.030	**.696**	.013
M7	-.096	**.647**	.254
M8	**.838**	.043	-.009
M9	**.983**	-.297	.007
M10	.062	.408	.425
M11	-.221	.429	**.553**
M12	-.097	**.916**	-.106
M13	.471	.487	-.481
M14	**.509**	.302	.076
Initial Eigenvalue	5.282	1.679	1.097
Rotation Sums of Squared Loadings	3.977	4.169	3.249
Cumulative variance explained		61.981	

Thus, the items M5, M10 and M13 were successively excluded from the analysis. After that, the analysis still revealed a three-factor solution (Table 19). Also, the required conditions concerning KMO (=0.791) and Bartlett test of sphericity (=0.000) underlined a valid variable set. All variables had sufficiently high MSA values and communalities (both requiring values greater than 0.5) and the cumulative variance explained had increased to 66.083.

Table 19: Factor loadings of motivations – after deletion of loadings < 0.5

			Pattern Matrix Factor		
Item	MSA	Communalities	1	2	3
M1	.623	.807	.009	-.191	**.969**
M3	.755	.628	.060	-.033	**.787**
M4	.853	.674	**.650**	.146	.195
M6	.826	.619	.112	**.788**	-.141
M7	.834	.539	-.023	**.530**	.332
M8	.766	.741	**.830**	.090	-.041
M9	.736	.772	**.964**	-.184	-.088
M11	.832	.581	-.157	**.531**	.426
M12	.829	.733	-.006	**.931**	-.192
M14	.843	.514	**.553**	.177	.136
Initial Eigenvalue			4.105	1.440	1.064
Rotation Sums of Squared Loadings			3.118	3.147	2.711
Cumulative variance explained				66.083	

Further, no significant cross-loadings were present in the emerged factor solution. Cross-loadings describe cases where a variable significantly loads on several factors. The EFA process aims at reducing the occurrence of such cross-loadings to facilitate accurate interpretation of the factor solutions (Hair et al., 2014, p. 117).

Table 20: Final set of motivational factors (ordered by loadings)

Item		Factor 1	Factor 2	Factor 3
M9	Competition from other C2C certified companies	.964		
M8	Potential loss of market share if the company does not implement C2C	.830		
M4	Expectation to contribute to the company's cost reduction	.650		
M14	Expectation to better comply with current legislation	.553		
M12	Expectation to improve customer satisfaction		.931	
M6	Expectation to increase sales		.788	
M11	Expectation to be perceived as a market leader		.531	
M7	Expectation of C2C to be a source of new opportunities		.530	
M1	Fit of C2C with the company's philosophy			.969
M3	Company CEO's/top management's strong endeavor to achieve C2C certification			.787

Table 20 presents the final interpretable factors and ordered loadings linked to the included variables, which formed the basis for the labelling and interpretation (KMO = 0.791, significance of the Bartlett test of sphericity = 0.000).

The factor labelling process should take into account that variables with higher loadings represent the core characteristics of the factor more strongly and should therefore be particularly reflected by the factor name (Hair et al., 2014).

For the first factor, this applies predominantly to the variables M9 and M8 which describe a motivational trigger mainly based on the concern of competition or loss of market share. This indicates a perceived market pressure that C2C could be a response to. The other two variables underline this context as they refer to motivations linked to cost reduction and legal compliance. Thus, the first motivational factor is labelled 'competitive pressure'. The second factor solution has a very high loading from the variable M12 presenting an expected improvement in customer satisfaction. Supported by two further variables linked to an increase of sales and an ameliorated market positioning, the factor suggests that the underlying motivations are related to perceived opportunities for the business, so that the factor is named 'expected benefits'. The third factor comprises of an item pointing at the initiation of C2C being triggered by the fit of the concept with the company's philosophy as well as a variable on the critical role of the top management in engaging in C2C implementation. Therefore, it can be concluded that the last factor represents a motivation that mainly stems from a standpoint considering the company's philosophy and that is present on the agenda of decision-makers. That in turn suggests a high strategic relevance, hence the factor is called 'strategic decision'.

Based on the aforementioned lack of clear-cut thresholds (chapter 4.4.1), an accurate interpretation of the factors can be reinforced by the application of further reliability measures (Stevens, 2002).

Reliability of motivational factors

Two well-established and common measures were applied to ensure the reliability of the identified factors and the underlying constructs of variables: the Cronbach's alpha measure and the inter-item correlation (IIC). To also assess the correlation of an item to the total scale score, the corrected item-to-total correlation (CITC) is taken into account. This measure complements the reliability assessment and is particularly suitable for a smaller size of indicators as it excludes the variable itself from the overall summation and thus leads to a clearer assessment of the correlation of each single variable to the identified factor (Nunnally and Bernstein, 1994). Cronbach's alpha (a) describes an item interrelatedness, often referred to as internal consistency. It ranges between 0 and 1, with higher values representing higher reliability of the construct (Cronbach, 1951). There are controversial debates about a required clearcut threshold. In the present study, due to its exploratory character and the lack of pre-existing empirical validation in this research area, the analysis follows the recommendation of Nunnally and Bernstein (1994) and Robinson et al. (1991) to apply the minimum threshold of 0.60 for the reliability coefficient.

The inter-item correlation measures the average correlations among all items included in the factor and should be greater than 0.3 (Robinson et al., 1991). Less univocally, the recommendations for CITC vary in a range of minimum values from 0.3. to 0.5 depending on different research scholars and settings (Ladhari, 2010). Besides a suggested cut-off value of 0.5 indicated in common statistics literature (e.g. Hair et al., 2014), the range of potential cut-off points also calls for a holistic consideration of several aspects before excluding a variable from the factor solution. For example, the research of Wolfinbarger and Gilly (2003) who suggest the application of a minimum CITC value of 0.4 if the factor loadings are adequately high (> 0.5) for the examined construct and adequately low (no loading of > 0.50 on two factors) on the other factors. The selection of the final set of items per factor would thus need to take various criteria into account. Table 21 shows the results of the reliability tests for the final factor solutions on motivations.

The reliability values are in accordance with the formulated reliability requirements (see Appendix 8.3 for the inter-item correlation tables), thus no further items were excluded from the factor solutions[57].

Table 21: Results for Cronbach's alpha and CITC for motivational factors

		Cronbach α	CITC
Competitive pressure		.812	
M9	Competition from other C2C certified companies		.615
M8	Potential loss of market share if the company does not implement C2C		.711
M4	Expectation to contribute to the company's cost reduction		.656
M14	Expectation to better comply with current legislation		.545
Expected benefits		.739	
M12	Expectation to improve customer satisfaction		.574
M6	Expectation to increase sales		.534
M11	Expectation to be perceived as a market leader		.490
M7	Expectation of C2C to be a source of new opportunities		.524
Strategic decision		.659	
M1	Fit of C2C with the company's philosophy		.528
M3	Company CEO's/top management's strong endeavor to achieve C2C certification		.528

Organizational context factors

Building on the data preparation measures, not all of the initially collected organizational context variables were included in the EFA. After excluding items due to missing values, a further elimination of variables resulted from the assessment of the necessary requirements with respect to MSA values and communalities greater than 0.5. Following these cut-off values, another six variables[58] were removed from

---

57 Only in the case of M11 for the factor 'expected benefits', the CITC value is slightly below the commonly applied cutoff value of 0.5. However, as the deviation is minimal and the factor loadings adhere to the specified requirements of Wolfinbarger and Gilly (2003), the variable was not excluded from the construct.

58 Due to "unacceptable" MSA values, the variables TS4, CC4, PAR5 and PAR6 were deleted. The variables CC2 and PAR7 were excluded based on insufficiently high communalities.

Exploratory Factor Analysis

the set which formed the basis for the EFA. After that, the remaining 12 variables fulfilled the requested criteria regarding MSA values and communalities (Table 22).

Table 22: MSA and communalities for organizational context variables – after deletion of TS4, CC2 and CC4

Item		MSA	Communalities
IMP1	The company has integrated C2C in procedures and work instructions	.840	.723
IMP2	The company has identified specific persons and positions responsible for C2C implementation	.636	.687
IMP3	The company keeps records of the training provided to staff in relation to the implementation of C2C	.799	.566
IMP4	The company has obliged its supply base to supply according to C2C	.714	.843
IMP7	The extent to which you believe that at this point in time C2C philosophy, standards, and methods have been implemented throughout your company	.822	.569
PAR1	We feel indebted to EPEA and/or MBDC for what they have done for us	.751	.769
PAR2	The company's employees share close social relations with the employees from EPEA and/or MBDC	.797	.719
PAR3	Our relationship with EPEA and/or MBDC can be defined as "mutually gratifying"	.782	.753
PAR4	We expect that we will be working with EPEA and/or MBDC far into the future	.696	.854
TS5	In the course of implementing C2C, the company made fundamental changes to the existing product and processes	.720	.785
TS6	The innovation process had to be significantly adapted to C2C	.755	.815
TS7	C2C implementation had a significant impact on the entire value chain of the company	.833	.731
Kaiser-Meyer-Olkin Measure of Sampling Adequacy		.767	
Significance of Bartlett's Test of Sphericity		.000	

Hereafter, the loadings for the extracted factor solutions were examined. Four factors were identified, each of these with an eigenvalue greater than 1, significant loadings of all variables except of IMP3, which was excluded from the construct, and a cumulative variance explained of 73.448, which is greater than the recommended threshold of 0.6 (Table 23).

Table 23: Factor loadings of organizational context variables – before deletion

Item	Pattern Matrix Factor			
	1	2	3	4
IMP1	-.039	**.569**	.275	.271
IMP2	-.023	**.879**	-.017	-.174
IMP3	-.148	.282	.467	.289
IMP4	-.181	.041	.046	**.946**
IMP7	.043	**.583**	-.076	.343
PAR1	**.860**	.048	.228	-.205
PAR2	**.512**	**.665**	-.083	-.177
PAR3	**.795**	.052	.164	.009
PAR4	**.673**	-.049	-.387	.494
TS5	.012	.070	**.862**	-.022
TS6	.347	-.226	**.796**	.088
TS7	.182	-.159	.201	**.723**
Initial Eigenvalue	4.892	1.524	1.319	1.078
Rotation Sums of Squared Loadings	3.161	2.996	2.958	3.222
Cumulative variance explained	73.448			

Besides the loading being smaller than 0.5 of the variable IMP3, the variable PAR2 ('The company's employees share close social relations with the employees from EPEA and/or MBDC') shows a cross-loading which calls for a more accurate examination of the variable. Since the item PAR2 was part of an empirically validated construct on relational embeddedness (Rindfleisch and Moorman, 2001), the loading on the first factor underlines the consistency of the construct, as it comes along with the other items from this construct, PAR1, PAR3, and PAR4. This would suggest including the variable with the first factor. Nonetheless, the loading on the second factor would be plausible as well since the employees are a critical part in the context of the internalization of new practices. This would allow the inclusion of the variable in the second factor, which comprises other variables on the extensiveness of the implementation. Thus, the item could not be clearly excluded or assigned to one factor after the contextual evaluation. Therefore, the additional reliability measure *average variance extracted* (AVE), widely used for confirmatory factor analysis, was applied to check the convergent validity of the construct. The AVE is based on the communality values and indicates the amount of variance of an item that is explained by the extracted factor, which leads to a better understanding of the convergent validity. It is recommended that the AVE should be greater than 0.5 for a construct (Hair et al., 2014). In the present case, the AVE adhered to this cut-off value only when matched to the first factor (AVE = 0.549) while it was below when assigned to the second factor (AVE = 0.480). Consequently, both the contextual and analytical assessment supported the affiliation of the item in question to the construct of relational embeddedness. Since the resulting factor also met the necessary reliability requirements (see Table 26), the item remained part of the EFA and was included in the first factor for the following analyses[59] (Table 24). For all factors, the analytical assessment was complemented by a graphical evaluation based on the scree plots.

---

59 If the contextual or analytical assessment does not allow for an unambiguous assignment of the item to one factor, the item could also be excluded from the analysis in order not to distort interpretation of results (Hair et al., 2014).

Table 24: Factor loadings of organizational context – after deletion of loadings < 0.5

Item	MSA	Communalities	Pattern Matrix Factor			
			1	2	3	4
IMP1	.800	.742	-.061	**.610**	.235	.304
IMP2	.632	.668	.005	**.872**	-.137	-.093
IMP4	.757	.825	-.186	.121	**.911**	.052
IMP7	.856	.626	-.030	**.642**	.324	-.002
PAR1	.739	.776	**.869**	.010	-.207	.237
PAR2	.792	.713	**.533**	.614	-.171	-.053
PAR3	.796	.766	**.827**	.031	.010	.108
PAR4	.724	.843	**.686**	-.038	.464	-.374
TS5	.681	.848	-.050	.109	.006	**.888**
TS6	.730	.810	.312	-.215	.165	**.758**
TS7	.796	.795	.125	-.120	**.814**	.135
Initial Eigenvalue			4.606	1.433	1.296	1.078
Rotation Sums of Squared Loadings			3.227	2.838	3.036	2.525
Cumulative variance explained			76.478			

To facilitate the labelling process of the factors, all loadings were ordered corresponding to the extracted factor (Table 25). The variables of the first factor clearly indicate that the underlying construct of relational embeddedness, which was applied based on existing empirical evidence (Rindfleisch and Moorman, 2001) proved applicable also in the present context. As the construct was focused on the company's collaboration experience during C2C certification, the variables specifically addressed the standard-setting organizations EPEA and MBDC. Thus, the first factor can be labelled as 'relationship with certification partner'.

Table 25: Final set of contextual factors (ordered by loadings)

Item		Factor 1	Factor 2	Factor 3	Factor 4
PAR1	We feel indebted to EPEA and/or MBDC for what they have done for us	.874			
PAR3	Our relationship with EPEA and/or MBDC can be defined as "mutually gratifying"	.869			
PAR4	We expect that we will be working with EPEA and/or MBDC far into the future	.716			
PAR2	The company's employees share close social relations with the employees from EPEA and/or MBDC	.533			
IMP2	The company has identified specific persons and positions responsible for C2C implementation		.872		
IMP7	The extent to which you believe that at this point in time C2C philosophy, standards, and methods have been implemented throughout your company		.642		
IMP1	The company has integrated C2C in procedures and work instructions		.610		
IMP4	The company has obliged its supply base to supply according to C2C			.911	
TS7	C2C implementation had a significant impact on the entire value chain of the company			.814	
TS5	In the course of implementing C2C, the company made fundamental changes to the existing product and processes				.888
TS6	The innovation process had to be significantly adapted to C2C				.758

The second factor embraces variables, which address how extensively the company realised the implementation and which measures have been taken in order to adopt C2C practices. The factor loading of IMP2 being the highest emphasizes the importance of clear responsibilities for the C2C certification and is supported by the other variables, which both reflect thorough implementation steps throughout the company and its processes. Hence, the second factor is labelled 'C2C anchorage'. The third factor is dominated by the variable which describes the obligation of suppliers to C2C-conform supply and is complemented by a variable affirming the consequences on the value chain. Both items emphasize the necessity of integrating activities related to sourcing and value creation during C2C implementation and both loadings are very high. Hereafter, the factor is labelled 'integration into supply and value chain'.

The last factor refers to necessary adaptations of processes or products in the course of C2C implementation with two relatively high loadings from the variables TS5 and TS6, hence it is described as 'degree of change' in the following analyses.

Reliability of organizational context factors

Based on the previously described reliability measures, the factor solution for the organizational context was also examined with respect to inter-item correlation (see Appendix 8.3), corrected item-to-total correlation (CITC) and Cronbach's alpha (Table 26). All values were within the recommended ranges so that the reliability of the constructs could be assumed.

Table 26: Results for Cronbach's alpha and CITC for organizational context factors

Factor		Cronbach α	CITC
Relationship with certification partner		.796	
PAR1	We feel indebted to EPEA and/or MBDC for what they have done for us		.661
PAR3	Our relationship with EPEA and/or MBDC can be defined as "mutually gratifying"		.705
PAR4	We expect that we will be working with EPEA and/or MBDC far into the future		.563
PAR2	The company's employees share close social relations with the employees from EPEA and/or MBDC		.515
C2C anchorage		.702	
IMP2	The company has identified specific persons and positions responsible for C2C implementation		.422
IMP7	The extent to which you believe that at this point in time C2C philosophy, standards, and methods have been implemented throughout your company		.583
IMP1	The company has integrated C2C in procedures and work instructions		.615
Integration into supply and value chain		.768	
IMP4	The company has obliged its supply base to supply according to C2C		.624
TS7	C2C implementation had a significant impact on the entire value chain of the company		.624
Degree of change		.758	
TS5	In the course of implementing C2C, the company made fundamental changes to the existing product and processes		.610
TS6	The innovation process had to be significantly adapted to C2C		.610

Satisfaction with C2C implementation (DV)

As the initial research framework defined 'satisfaction' as dependent variable, no values have been replaced for the items related to this measure. To ensure the validity and reliability of the construct, which was composed of variables from previous research and the qualitative study, the items related to the construct were included in the scope of the EFA. Similar to the previous analyses, the EFA was conducted as Principal Component Analysis with Promax Rotation. From the initially seven items comprised in the survey, two had to be excluded due to missing data over 10 %. The remaining items were included in the EFA and confirmed a one-factor solution[60] (Table 27).

Table 27: Factor loadings of satisfaction-related variables

Item	MSA	Communalities	Factor 1
S2	.810	.734	.857
S4	.837	.702	.838
S1	.838	.692	.832
S7	.831	.663	.814
S3	.893	.584	.764
Initial Eigenvalue			3.375
Cumulative variance explained			67.493
Kaiser-Meyer-Olkin Measure of Sampling Adequacy			.839
Significance of Bartlett's Test of Sphericity			.000

The reliability of the factor is supported by a Cronbach's alpha of 0.871 and CITC values greater than 0.5 (Table 28). Consequently, the construct could be included as the DV in the following regression analysis, which is the focus of the next chapter.

Table 28: Results for Cronbach's alpha and CITC for satisfaction (DV)

Factor		Cronbach α	CITC
Satisfaction with C2C implementation (DV)		.871	
S2	We are pleased to be associated with C2C		.758
S4	The company's image has improved because of the C2C implementation		.721
S1	We would recommend the implementation of C2C to our business partners		.725
S7	Please indicate your company's overall satisfaction with the C2C implementation		.700
S3	The implementation of C2C spurred innovation in the company		.635

---

60   As only one component was extracted, the solution cannot be rotated.

Control variables

Since three of the four control variables were measured as single-item variables and only one variable ('new product success') consisted of five items, no EFA was conducted for the covariates. However, in order to verify the reliability, the construct was examined before proceeding with the analysis. On the one hand, this led to the exclusion of two items in the course of the data cleansing process due to missing values greater than 10%. On the other hand, examining the construct for Cronbach's alpha and the CITC resulted in robust results within the recommended threshold (see Table 29). Therefore, the construct was included in the further analysis as a control variable.

Table 29: Reliability of control variable 'new product success'

Factor		Cronbach α	CITC
New product success		.856	
After the implementation of C2C standards, the company's C2C certified product(s)…			
NPS1	Overall, met or exceeded sales expectations		.837
NPS4	Met or exceeded market share expectations		.817
NPS5	Met or exceeded customer expectations		.555

4.5 Multiple linear regression

4.5.1 Research framework

Due to the lack of empirically tested theories and constructs, the underlying conceptual research framework stipulated a quantitative exploration of latent variables by applying a multiple linear regression model (Hair et al., 2014). Based on the potential influencing factors elaborated in the previous chapters[61], it is assumed that motivational factors as well as organizational context (independent variables, IV) have an impact on a company's level of satisfaction after going through the process of C2C implementation and certification (dependent variable, DV). Hereby, the results of the factor analysis (see chapter 4.4) formed the basis for estimating the DV based on the three motivational and four organizational factors. In order to capture potential effects resulting from a company's organizational complexity, the company size was

---

61 See chapter 4.1.1 for the comprehensive variable overview.

included as a moderator (see chapter 4.1.1). In addition to the direct effect on the dependent variable, the interaction effects with the factors of the organizational context were included[62]. The model controls for effects resulting from a company's current certification status, the company age, the most frequent C2C certification level and the success of the introduced C2C products (see Figure 23).

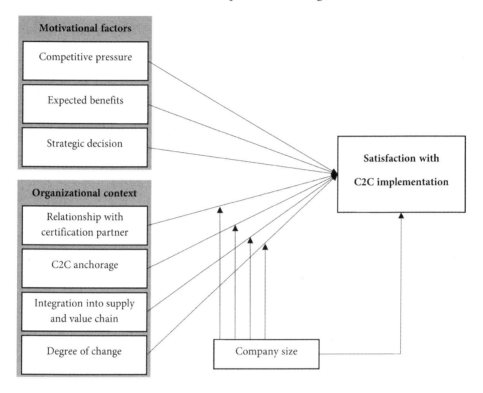

Figure 23: Research framework[63]

4.5.2 Underlying assumptions for the multiple regression analysis

Sample size for regression analysis
The statistical power of a regression analysis is closely linked with the sample size in the data set. Thus, there are various measures related to the number of observa-

---

62 In an earlier stage of the research project, the analysis has been conducted without testing for moderation effects. The preliminary results were presented at the R&D Management Conference 2016. Subsequently, the model has been adapted and further developed, thus the present results slightly differ from those preliminary, after deeper investigation and further analysis.

63 Author's illustration.

tions that should be evaluated before the regression analysis can be conducted and appropriately interpreted. One concrete rule of thumb provided by Hair et al. (2014) is a required minimum sample size of 50 and a preferred number of 100. A related indicator is the ratio of observations to variables, which is recommended to be at least 5:1, again preferably with a higher ratio such as 15:1 (Hair et al., 2014). At the same time however, a sample that is too large can lead to inaccurate interpretations due to significant relationships caused mainly by the large sample size per se. Therefore, other indicators should also be taken into account when evaluating the data sample size. One additional measure in this context is R^2, the coefficient of determination, which presents the share of the dependent variable's variation that is collectively explained by the independent variables. It can range between 0 and 100 % and the desired result is to achieve the highest possible value, implying that the selected regression model adequately fits the data (Hair et al., 2014). Another indication linked to the sample size and being critical for the result generalizability is the number of degrees of freedom (dF) in the regression model. They present the difference of the number of observations (sample size) and the number of estimated parameters and thus allows for an assessment of the prediction accuracy of the regression model (Hair et al., 2014). For the present data set, the sample size and the ratio of observations to variables meet the minimum requirements[64]. With respect to R^2 and dF, the values will be evaluated in the following after performing the multiple regression analysis.

Testing assumptions of multiple linear regression
In the following section, the data is examined with regard to the requirements of *linearity, normality, homoscedasticity, autocorrelation of the residuals,* and *multicollinearity* before proceeding with the regression analysis. In a first step, the basic assumption of a multiple linear regression, namely the linear relationship between dependent and independent variables, is examined. For this purpose, a visual inspection of the normal P-P plot, which is a scatterplot displaying the residuals, helps to evaluate the relationship of the predictor variables with the outcome variable. The generated scatterplots indicate a linear relationship, thus it was presumed that the linearity requirements are met with the present data set.

A further requirement, often referred to as necessary when conducting multiple regression is the normal distribution of data (Backhaus et al., 2016; Cohen et al., 2003; Hair et al., 2014). In the case of smaller sample sizes the focus of the normality analysis should lie on the residuals' distribution (Cohen et al., 2003). Hence, in order to identify potential nonnormality of residuals before proceeding with the regression analysis, two commonly applied tests were conducted, the Kolmogorov-Smirnov nor-

---

64 See chapter 4.3 and 4.4.

mality test and the Shapiro Wilk test (Hair et al., 2014). In both tests, the null hypothesis suggests that the data is normally distributed, hence a significant result indicates the absence of data normality. In the present case, both tests were not significant on the 5% significance level which suggests that there is no significant deviation from normality for the residuals (Kolmogorov-Smirnov $p = 0.200$ and Shapiro-Wilk $p = 0.225$). In addition to the analytical tests, the normality was also checked through a visual inspection of the skewness and kurtosis plots and confirmed a normal distribution of the residuals.

In a third step, the equality of the residuals' variance (homoscedasticity) in the data set was checked, which can be performed through both, a visual inspection of the scatterplots and an analytic assessment (Cohen et al., 2003; Hair et al., 2014). With respect to the shape of the scatterplots, the typical patterns of heterogeneous variances of residuals are triangle- or diamond shaped (Hair et al., 2014; Stevens, 2002). However, the present data set doesn't show such a pattern (see Figure 24), so that homoscedasticity can be assumed and the necessary requirement for a multiple regression is met by the present data set.

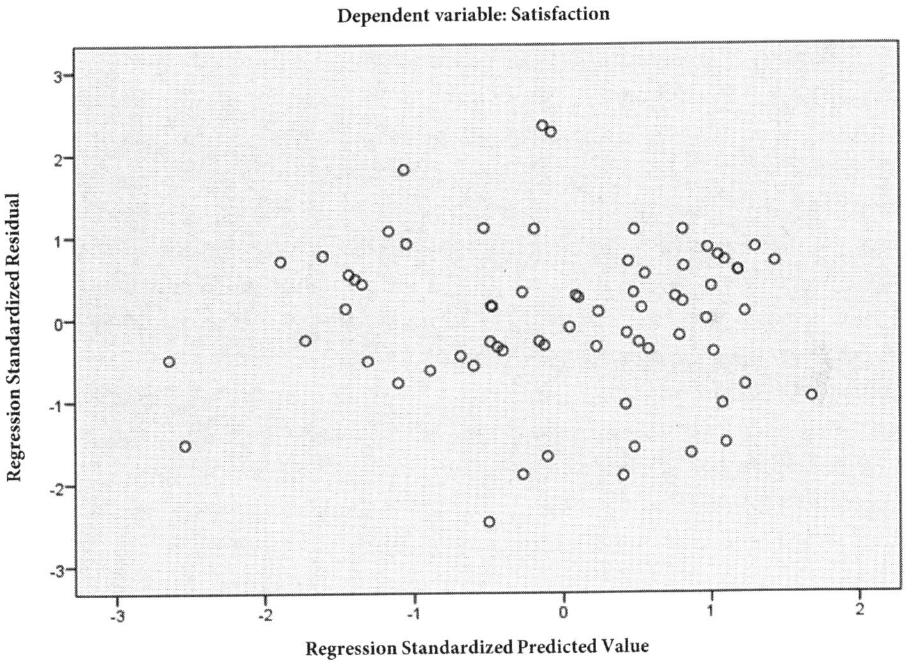

Figure 24: Scatterplot to test for homogeneity of residuals

This assumption was further tested with the help of a commonly applied statistical test, the Breusch-Pagan test (Breusch and Pagan, 1979; Cohen et al., 2003). The Breusch-Pagan test is based on the null hypothesis of homoscedasticity and should only be applied when residuals are normally distributed, which has been validated already. For the underlying data, the significance of 0.630 shows that there is no significant deviation from the assumption of homoscedasticity, thus, a homogenous variance of the residuals can be presumed.

Further, the data was checked for the *autocorrelation of the residuals* as the presence of positive or negative autocorrelations can lead to incorrect standard errors and confidence intervals (Cohen et al., 2003). The Durbin-Watson test, even though particularly suited for the evaluation of temporal series of observations, is an accepted measure in such a context and indicates the presence or absence of autocorrelation for the error terms. The values can range between 0, indicating a positive autocorrelation, and 4, which would suggest a negative autocorrelation. A value of d = 2 is considered to portend the absence of a dependency of the observations (Backhaus et al., 2016; Cohen et al., 2003). Performing the test for the underlying residuals results in a value of 1.921, thus it can be assumed that there are no autocorrelation issues for the following regression analysis.

Finally, one common and necessary analysis when conducting the multiple regression analysis is the check for *multicollinearity*, which reveals the correlation of independent variables among themselves. The aim is to increase the possible prediction by excluding those independent variables that are highly correlated with other independent variables as this indicates a critical amount of shared variance. As such, high levels of multicollinearity reflect higher standard errors, lead to incorrect estimations of coefficients and impede an accurate interpretation of the regression results (Cohen et al., 2003; Hair et al., 2014). A common measure to test for multicollinearity is the variance inflation factor (VIF), which presents the inverse of the tolerance value ($1-R^2$) and is recommended to range between 0.1 and 10 (Cohen et al., 2003; Hair et al., 2014). The present sample yielded VIF values which are all in the accepted range, with a VIF of 1.198 being the lowest and a VIF of 2.324 being the maximum value (see Table 30). Thus, no variable needed to be excluded from the analysis and the absence of multicollinearity can be assumed.

Table 30: Analysis of multicollinearity

Variables	Tolerance	VIF
Competitive pressure	.558	1.791
Expected benefits	.588	1.702
Strategic decision	.507	1.972
Relationship with certification partner	.605	1.652
C2C anchorage	.430	2.324
Integration into supply and value chain	.479	2.087
Degree of change	.584	1.711
Current certification status	.823	1.216
Age of the company	.834	1.198
Most frequent C2C certification level	.711	1.407
New product success	.716	1.397

Bias treatment

Since quantitative data collection usually represents a more or less random sample of a population, it can be subject to different bias effects, so that the underlying data set should be assessed with respect to these issues before further proceeding with the regression analysis (Weiber and Mühlhaus, 2014). Two widely established bias estimations address the differences between respondents and non-respondents (non-response bias) as well as the applied methodology (common method bias). In addition, taking into consideration that the survey was sent out to one company representative who replied on behalf of the company, the data set was also checked for a potential informant bias.

Non-response bias

To address a potential bias effect suggesting that only particularly satisfied or particularly unsatisfied companies participated in the survey, the non-response bias is a suitable measure to estimate whether the group of respondents would show a different response pattern than non-respondents. For this purpose, theory suggests splitting the sample in early and late respondents based on the assumption that the response behaviour of late respondents resembles the one of non-responders (Armstrong and Overton, 1977). Performing two-tailed t-tests to compare the two samples resulted in non-significant results (level of significance 5%) for all included variables. Thus it can be assumed that the underlying data is not biased with respect to non-respondents (Armstrong and Overton, 1977).

Common method bias

The criticality of the measurement error resulting from a common method bias has particular relevance in business research as this research field most commonly makes use of self-reported data (Doty and Glick, 1998; Fuller et al., 2016). This bias effect can occur when "the measurement technique introduced systematic variance into the measure" (Doty and Glick, 1998, p. 374). Yet, discussion among scholars is somewhat controversial, also because of the many potential causes for common method variance (Fuller et al., 2016; Podsakoff et al., 2012; Spector, 2006). For instance, Spector (2006) criticizes the theoretical debate as being oversimplified and the phenomenon having become a sort of 'urban legend'. Despite a potential misinterpretation, Spector concludes that the underlying assumptions of such a bias effect hold true and hence encourage a differentiated perspective towards common method bias, e. g. such as suggested by Podsakoff et al. (2003). This means that effects can – but do not necessarily have to - exist, and that these effects can vary depending on the research design and context. As extensively elaborated in the work of Podsakoff et al. (2003), there are numerous potential sources for common method bias, many of which are rooted in the formulation and setup of the survey in the first place. Thus, potential remedies can be clustered in preventive (or procedural) measures, which can be addressed ex ante through an adequate survey design, and statistical tests subsequent to the data collection. The preventive measures were taken into account during the research design and survey development. In the following, the analysis intends to verify in how far the present data set is still subject to biasing effects.

In a first step, since the research context didn't allow for a data collection from various sources, the survey design presents a source for potential biasing issues. As suggested by Podsakoff et al. (2003), a "temporal, proximal, psychological, or methodological separation of measurement" could be a suitable remedy to more explicitly decouple the predictor and the criterion variables (Podsakoff et al., 2003, p. 887). The present research setup made such a separation difficult, so that more emphasis was put on the communication with respect to data handling towards the respondents. Therefore, in the beginning of the survey, the participants were guaranteed a confidential data handling and their anonymity. Further, the introductory text made explicit that the survey was launched in the course of a research project, independent of C2C-related organizations, e.g. the EPEA, so that respondents would not feel restraint to also report on negative experiences. Finally, it was highlighted that there is no right or wrong answers but that respondents should reply to the best of their knowledge. These measures should minimize potential biased response patterns resulting from participants anticipating a certain response behaviour or providing socially desirable responses. In addition, items and scales were framed clearly and pre-tested to prevent potential misunderstandings. To reduce a potential acquiescence bias, all included scales contained concrete verbal labels not only numerical values[65] (Podsakoff et al., 2012; Tourangeau et al., 2000).

In a second step, two analytical measures were applied to statistically test for potential common method bias effects. Yet, also such analytical methods are no less the subject of ambivalent discussions and must therefore be tailored to the specific research context (Podsakoff et al., 2003). For the present research design, Harman's single factor test presented itself as suitable as it provides insights on the variance that would be explained by only one factor (Fuller et al., 2016). The test requires the execution of an unrotated exploratory factor analysis from which one single factor should emerge accounting to a large share of the extracted variance in order to reveal a common method bias issue. Applied to the present data set, the EFA indicated six factors in total, with the first factor accounting for 33.58 % of the extracted variance. Even though there is no clear recommendation on a specific cut-off point[66], this value cannot be considered as a substantial share (Fuller et al., 2016; Podsakoff et al., 2003).

The second analytical measures investigates the correlations of the factors. The correlation matrix provides a suitable means to further check for common method bias (Pavlou et al., 2007). For correlations to be regarded as critical and subsequently excluded, the constructs should correlate with a value of more than 0.9. For the present data set, no correlation coefficient for the included variables at hand exceeded this value, the highest correlation being r = 0.686. Hence, this test neither suggests the presence of a biasing issue due to common method variance.

Informant bias

Scholars have debated over the adequate number of necessary data sources per company for decades or over the suitability of single informants when collecting organizational data for quantitative research projects (Golden, 1992; Seidler, 1974). On the one hand, the single informant method is a commonly applied method as it facilitates data collection. On the other hand, it also involves some caveats regarding the objectivity of the respondents and potentially inaccurate recalling of past events (Golden, 1992; Kumar et al., 1993). With respect to the assessment of the surveyed

---

65 Chapter 4.1.2 provides a detailed elaboration on the survey development.
66 Numerous studies refer to a cut-off value of at least 50 % for the common method variance to be considered as problematic. A comprehensive review of studies having applied Harman's single factor test can be found in Fuller et al. (2016).

topics in question, the tenure or organizational position of the respondent may play a critical role (Golden, 1992; Seidler, 1974). Therefore, the data set was checked for a potential informant bias effect by splitting the respondents according to their position level, which was part of the survey questions. In order to achieve a maximum spread, the sample of respondents who indicated their position level as 'employees' were contrasted with the respondents having stated they are from the 'top management'. Then, a two-tailed t-test was conducted for all variables included (significance level of 5%). The test revealed no significant results except for the motivational factor 'expected benefits' with a significance of 0.017, which seems to be mainly driven by the significance of the underlying item 'The expectation of C2C to be a source of new opportunities' (M7). Since all other variables did not indicate significant divergence of the response pattern, it is assumed that a key informant bias is not a critical issue for the present data sample.

After verifying that the critical underlying assumptions are respected in the present data set, the multiple regression model can be estimated. The results are elaborated in the following chapter.

4.5.3 Results of multiple regression analysis

Overall results and model summary

The multiple regression analysis was conducted based on the idea of a hierarchical (or sequential) regression, which implies that independent variables are included in the regression model based on an upfront selection. This procedure, unlike the stepwise forward or backward selection (Stevens, 2002), is based on a 'causal priority' for the introduction of variables (Cohen et al., 2003). In the present analysis, the baseline model (model 1) consists of four control variables, including the factor 'new product success' which seemed an obvious predictor[67] for a company's satisfaction (DV) and hence a suitable foundation for checking whether additional variables will improve the variance explained. In a second step, the motivational and organizational factors were added as a next block to the analysis (model 2), followed by the moderating variable 'company size' and the respective interaction effects (model 3). The hierarchical approach can thus support a clearer interpretation of the variables' role in the explained variance as it underlines the respective increments in R^2 (Cohen et al., 2003). A summary of the overall model results is shown in Table 31.

---

67 See chapter 4.1.1.

Table 31: Model summary and regression results

	Model 1		Model 2		Model 3	
	b	Sig	b	Sig	b	Sig
Current certification status	-.482	.059	-.205	.268	-.232	.218
Age of the company	.019	.733	-.012	.756	-.052	.272
Most frequent C2C certification level	.181	.075	.037	.628	.002	.982
New product success	.345	.000	.227	.001	.264	.000
Competitive pressure			-.190	.007	-.208	.004
Expected benefits			.053	.569	.057	.544
Strategic decision			.169	.024	.212	.007
Relationship with certification partner			.364	.000	.420	.000
C2C anchorage			-.100	.359	-.198	.079
Integration into supply and value chain			.189	.005	.205	.005
Degree of change			.101	.107	.046	.467
Company Size					-.057	.124
Interaction (Relationship x Size)					-.085	.340
Interaction (Integration x Size)					.200	.019
Interaction (Anchorage x Size)					-.063	.459
Interaction (Change x Size)					-.002	.971
R^2	.353		.741		.781	
Adjusted R^2	.312		.690		.712	
R^2 Change	.353		.388		.040	
F Change	8.594		11.995		1.863	
Sig. F Change	.000		.000		.117	

Quantitative exploration of motivational factors and organizational enablers

Before elaborating the individual direct and indirect effects (Figure 25), the overall model measures are examined more closely. With respect to the variance explained, the included control variables account for a significant increase of R^2 and F in model 1 already, however only explain a minor part of the DV variance. The inclusion of the motivational and organizational factors in model 2 contributes a major increment of the model fit as the value for adjusted R^2 increased from $R^2 = 0.312$ to 0.690. This measure slightly further increases to a value of 0.712 after adding the moderating variables.

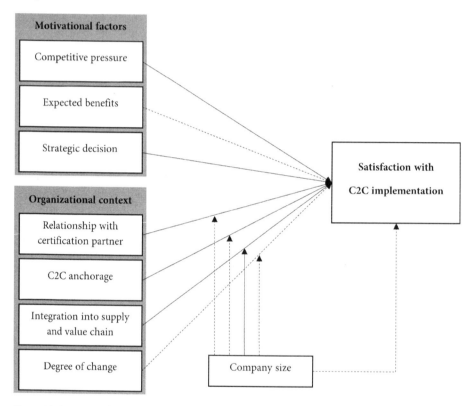

Figure 25: Overview of final results († $p < 0.1$; * $p < 0.05$; ** $p < 0.01$; *** $p < 0.001$)

Direct effects

The baseline model (model 1) controlled for the effects resulting from a company's current certification status, its age and most frequent C2C certification level as well as the success of the C2C certified product(s). The analysis only yields a significant effect for 'new product success', yet with a very high significance level of $p < 0.001$ and a regression coefficient of $b = 0.345$. The significance remains at this high level for model 2 and 3 and the regression coefficient slightly decreases to $b = 0.264$

(model 3). The covariates 'current certification status' and 'most frequent C2C certification level' show effects on a lower significance level ($p < 0.1$) in model 1, which dissipate in model 2 and 3.

With respect to the independent variables, five direct effects can be observed, two for the motivational factors and three for the organizational context. While the motivational factor 'competitive pressure' resulted in a negative effect ($b = -0.208$, $p = 0.004$), the variable 'strategic decision' positively influences the DV ($b = 0.212$, $p = 0.007$). When looking at the organizational context, most dominantly the variable 'relationship with certification partner' not only shows the highest significance level ($p = 0.000$) but also the highest coefficient ($b = 0.420$). The second highest regression coefficient ($b = 0.205$) can be observed for the 'integration into supply and value chain' with the significance value of $p = 0.005$. The third direct effect ($p = 0.079$) shows a negative coefficient ($b = -0.198$) for the factor 'C2C anchorage'. The independent variables 'expected benefits' and 'degree of change' don't significantly influence the dependent variable.

Moderating effects

Model 3 estimated the direct effect of 'company size' on the DV in addition to four potential interaction effects with the organizational context factors. As elaborated in chapter 4.1.1, this variable served as an indication for the organizational complexity. Hence, it was investigated in how far the company size moderates the effects of the organizational factors on the company satisfaction. This analysis can expedite the interpretation of the identified effect and "helps to establish the boundary conditions" of it (Hayes, 2013, p. 208). The interactions were calculated as the product term of the moderator and each of the four organizational factors (Hair et al., 2014; Hayes, 2013). The analysis yielded no significant impact for three out of four effects, nor did the analysis reveal a significant direct effect of company size on satisfaction. The one significant positive effect results from the interaction of 'integration into supply and value chain' and 'company size' ($p < 0.05$, $b = 0.200$). With respect to the overall regression analysis, the inclusion of the block with the interaction variables into the overall model led to a minor improvement of R^2 and adjusted R^2 and didn't show a significant F change.

Facilitating the interpretation, a simple slope analysis is an established means of visualizing the significant effect by plotting the predicted relationship between the dependent and independent variable at different values of the moderator variable (Dawson, 2014). The selection of these values can be based on specific contextual reasons or, if the particular context doesn't indicate an individual determination, as in the present case, the meaningful values are stipulated based on the standard deviation (SD) and the mean value (Cohen et al., 2003; Dawson, 2014). Following this approach, the simple slope analysis for the significant positive moderation effect is conducted based on a small company size, i. e. – 1 SD below the mean of the moderator, a large company size, + 1 SD above the mean respectively, and the mean value of the moderator for the medium company size. The results are illustrated in Figure 26.

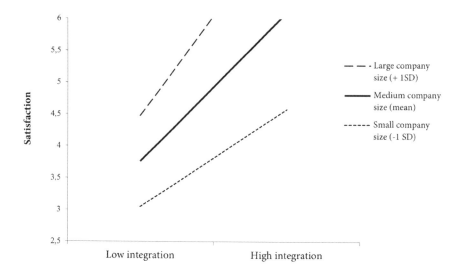

Figure 26: Simple slope analysis for significant moderation effect

To further probe into the moderating effect, the Johnson-Neyman technique was additionally applied in order to identify potential regions of significance (Hayes, 2013; Potthoff, 1964). For this reason, the significance of the moderation effect can be analysed more granularly for all values of the moderator variable. With the help of the PROCESS macro[68], developed by Andrew Hayes, the analysis can be conducted using the IBM SPSS Statistics software. One requirement for the application is that the moderator M presents a continuum, which is valid for the present moderator 'company size' (Hayes, 2013). Hence, the Johnson-Neyman technique can be applied for the present interaction. Supporting the visual inspection of the simple slopes, the analysis didn't indicate any "statistical significance transition points within the observed range of the moderator" (Hayes, 2013). Conclusively, the positive effect of 'integration into supply and value chain' on the company's satisfaction applies to companies of different sizes, however this effect diminishes with decreasing values of the moderating variable, i. e. the smaller the company is.

---

68   http://www.processmacro.org/index.html

5 Discussion of findings

The following chapter discusses the main results with respect to the critical driving forces for successful implementation of C2C, the case example for CE-related practices at hand, as well as organizational requirements for a prolific organizational environment. Hereby, section 5.1 focuses on research questions RQ1 and RQ2 by addressing motivational and organizational context factors which precede a satisfactory outcome with the C2C implementation. Moreover, section 5.2 outlines how an organizational environment would encourage a successful shift towards CE (RQ3).

5.1 Determinants of CE adoption

This chapter engages in a discussion based on findings from both the qualitative and quantitative exploration of motivational and organizational factors and aims at critically evaluating the developed propositions (chapter 3.3). Due to the lack of empirical evidence on the concrete example of C2C implementation, potential explanations and reasoning are complemented with supporting evidence from previous research on more general and sustainability-related activities. The evaluation of the developed propositions follows the work of Miles et al. (2014), who suggest different degrees of support, which are categorized in 'strong', 'qualified', 'neutral' and 'contradictory'

5.1.1 Driving forces for implementation

The first research question focused on the exploration of motivations and expectations of companies, which decided to adopt C2C and apply for the certification program. Even though different scholars have investigated the reasons for initiating sustainability-related activities (e.g. Bansal and Roth, 2000), the magnitude of such activities is wide, leading to a deficient comparability and hitherto making it difficult to derive clear patterns. This was also underlined by further publications dealing with this question in the years following the first study, e. g. by Schaltegger and Hörisch (2017).

Despite the challenging comparability between studies, two common tenets emerged in this academic discussion. On the one hand, scholars identified a more opportunity-driven motivation, for instance what is described as 'profit-seeking' by Schaltegger and Hörisch (2017) or 'economic opportunities' by Bansal and Roth (2000). Such motives are triggered by the expectation of improving the company's financial situation and sustainability-related innovations are regarded as a potential competitive advantage. On the other hand, scholars described a more threat-avoiding perspective, also referred to as 'legitimacy-seeking' (Schaltegger and Hörisch, 2017) or 'institutional pressure' (Chappin et al., 2015). In these cases, risk reduction and the response to different kinds of pressure, e. g. stakeholder or legislation, seem to be more dominant than an active pursuit for opportunity maximization (Crilly et al., 2012). Yet, it remained open whether and in how far these motives co-exist and if they can influence the actual implementation behaviour. In addition to these two streams, a range of other potential motivations has been analysed, for instance the role of individual decision-makers or ethical concerns (Bansal and Roth, 2000). To contribute to a clearer picture of dominant drivers, particularly in the context of Circular Economy activities, the basis for answering the first research question was built with the help of open questions in the course of an interview series (see chapter 3). This analysis pointed toward a wide range of motivations, also reflecting both opportunity-driven motives as well as threat-avoiding factors. Complemented with motives from extant theory, the list of motivations for C2C implementation was further scrutinized with the help of an exploratory factor analysis, resulting in three groups of main motives: *competitive pressure, expected benefits* and *strategic decision*.

It already became clear during the interview series that the motivations were not mutually exclusive in any of the cases. Still, in most of the cases, C2C efforts were based on one major trigger. The degree of support (Miles et al., 2014) for the developed proposition P2[69] can thus be regarded as strong.

The results also pointed out that depending on the initial motivation, the implementation and certification efforts come along with specific expectations, which in turn have an impact on how companies undergo the implementation process and how they ultimately evaluate these efforts. When companies primarily act out of pres-

---

69 P2: *The decision to engage with C2C can be triggered by different motivations, which range between more pro-active and opportunity-related motives to more reactive and threat-avoiding motives. Further, initiation can also emanate from decision-makers, often being aligned to the corporate strategy or goals. One single dominant trigger can hardly be isolated within this group of motives.* See chapter 3.3 for the detailed elaboration.

sure, this can influence the extent of implementation[70] as well as the satisfaction with the outcome, not only from the corporate perspective but also from a more individual employee viewpoint (Bansal and Roth, 2000).

Competitive pressure: Resulting from the EFA (chapter 4.4), the factor 'competitive pressure' included aspects such as better compliance with legislation, pressure from competitors or an anticipated loss of market share due to a shrinking customer base. This underlines that the range of potential sources of pressure which are exerted on companies can be diverse (Bansal and Clelland, 2004; Chappin et al., 2015; Crilly et al., 2012) and affect different levels of the organization. For instance, Schaltegger and Hörisch (2017) describe a positive effect of legitimacy-related motives on employee motivation, which would facilitate the implementation process.

The regression results of the above analysis showed that motives relating to pressure have a significant ($p < 0.01$) negative effect on the company's satisfaction. Taking into account that the response to pressure or competition is often a reactive action, it is reasonable to conclude that such activities might hamper a positive evaluation of the C2C implementation process. When companies try to escape from fines or decreasing revenues, such activities are seldom regarded as satisfactory or successful (Bansal and Roth, 2000; Chen, 2008). Further, the implementation efforts might be perceived as time critical leading to a rather superficial level of implementation, for instance by adapting minor product characteristics instead of focusing on a more holistic change process in the early phases of the innovation process. A shallower implementation might as well inhibit a re-design of the supply chain which was emphasized as a major success factor in prior research (Bocken et al., 2014; Ellen MacArthur Foundation, 2013; McDonough and Braungart, 2002a) and in the course of the above analysis. Thus, a short-term perspective towards the C2C implementation and certification can also result in dissatisfaction since the outcome doesn't correspond to the expected returns in the set period of time.

Bansal and Roth (2000) found in their study that competitively pressured activities are subject to a more critical cost-benefit assessment, hence the expectations with regard to the adopted practice are higher when it comes to returns and expected financial benefit. Such concerns and the balancing of cost and benefit were also one of the dominant topics in the course of the qualitative study. The particularly high financial expectations of competitively motivated companies seem to contribute to a lower satisfaction with C2C implementation and certification. Further, competitive pressure might be a more common motive for companies in certain industries or countries which are subject to new legislation or new consumer trends. Such an

---

70   See the recent publication of Smits et al. (2020) for a detailed elaboration of motives to adopt C2C and their effect on the implementation process.

environment can hamper a company's possibilities to use their C2C efforts as a differentiating feature as it becomes more of a standard for numerous companies and thus the implementation could rather cause discontent on the company side (Bansal and Roth, 2000; Schaltegger and Synnestvedt, 2002).

Expected benefits: Various previous studies have identified sustainability in general as a potential driver for innovation and competitive advantage (e.g. Esty and Porter, 1998; Hansen et al., 2009; Nidumolu et al., 2009; Porter and Kramer, 2006). Even though the investigated sustainability activities lack a clear comparability with regard to their actual form and CE activities go beyond the classic understanding of sustainability efforts, the study results point at certain characteristics that can foster innovative thinking, enhance re-design of products and processes or leverage resource use. As described in chapter 2.2.3, numerous scholars have underlined similar opportunities of CE-related innovations. Besides that, the concept of Circular Economy, in the present study operationalized through the C2C standard, further complements the list of potential benefits as it explicitly focuses on materials management and the elimination of waste as such by offering an alternative business and consumption model with respect to the use of finite resources (Ellen MacArthur Foundation, 2012; Ghisellini et al., 2016). Looking at the factor 'expected benefits' resulting from the exploratory factor analysis, companies seem to be aware of the opportunity to increase competitiveness, customer satisfaction and sales by engaging in C2C. Yet, the regression analysis did not reveal a significant effect on the company's satisfaction in contrast to the negative significant effect resulting from 'competitive pressure'. This absence of a significance might also be due to the relatively low variance of the factor (0.515) compared to the variance values of the other two motivational factors (0.971 for 'competitive pressure' and 0.928 for 'strategic decision'), indicating a stronger homogeneity of the data across companies. This would resonate with the idea that companies initiating CE activities most probably expect positive outcomes, even though this might not be a major driving force for the implementation itself.

Strategic decision: The third motivational factor 'strategic decision' positively influenced the satisfaction and included the role of the company's top management as well as the strategic fit of the C2C concept. The criticality of these aspects, not only in conjunction with sustainability-related activities but for innovation success more generally, has been pointed out by previous research as well (e.g. Koen et al., 2002; Tollin and Vej, 2012; Wheelwright and Clark, 1992). These studies identified, amongst other points, the crucial role of a 'champion' who promotes and enforces an implementation process on the corporate level as a success factor for implementing new practices. For the adoption of the C2C standard, this aspect seems to be of particular relevance as both the qualitative and quantitative study revealed a high need for the dedication of resources. Hence, internal ambassadors should have a certain decision power and budget responsibility.

Depending on the corporate strategy or business goals, such as addressing new customers or increasing productivity, the decision might also be a "side-effect" of business-related considerations of the management (Carrillo-Hermosilla et al., 2010). Not only with regard to sustainability-related activities (e.g. Basu and Palazzo, 2008), the importance of leadership and top management roles for determining a company's culture and business focus has been the subject of numerous research projects throughout the last decades (see e.g. Birkinshaw et al., 2008; Schein, 2010). Therefore, the importance of decision-makers who are shaping a corporate culture and strategy, often by balancing disjointed interests, seems critical in the process of shifting towards a Circular Economy. Such individuals or groups of decision-makers influence the substantial anchoring of a new practice, such as C2C, in the company's core processes and values (Bansal and Roth, 2000; Hahn et al., 2015). Depending on the company size and corporate culture, the management can either foster or hamper a more rigorous implementation process according to the applied leadership style, as described by Vaccaro et al. (2012) in their work on transformational and transactional leadership.

The above regression analysis also underlined that the strategic fit of C2C to the company's culture has a positive significant ($p < 0.01$) effect on whether the C2C implementation process is perceived as successful or not. This emphasizes the importance of carefully selecting the sustainability-related practice in order for it to match with the company's philosophy and to on-board and commit the management board or critical decision-makers. Being one of the most critical success factors when introducing change processes in a company, the formulation and communication of a vision and the active role of key players would hence foster the implementation process (Kotter, 1995). This could lead to a more satisfactory outcome than randomly implemented activities without the backing of internal ambassadors and clear embedding of an overarching vision. Furthermore, the factor 'strategic decision' implies a longer-term perspective, which favours smoother adaptations of processes, also including more time for capability building and training of staff involved in the C2C certification process (Ketata et al., 2015; Vaccaro et al., 2012). Also, with the right decision-makers in charge for the initiation as well as implementation, it is more likely that the company will remain on the implementation path. Thus, in the case of "ethical criteria" (Bansal and Roth, 2000, p. 728) being the decision basis, potential challenges or even financial losses might be disregarded and accepted.

Ultimately, the assumed relationship between motivations and the perceived satisfaction, as formulated in proposition P3[71], can be observed for two of the three motivational factors. Hence, the degree of support can be stated as 'qualified'. The presence of an interconnectedness of the motives makes it difficult to evaluate one dominant driver for satisfaction.

5.1.2 Organizational context and the implementation process

Alongside the importance of motivational factors, the organizational context was in the research focus in order to answer research question RQ2 with regard to critical organizational factors and their effect on a company's C2C implementation. As elaborated in chapter 4.4, the exploratory factor analysis resulted in four organizational factors and chapter 4.5 estimated how they influence the perceived satisfaction with C2C.

Relationship with certification partner: Looking at the result of the regression model, the variable 'relationship with certification partner' accounted for the highest effect size with the highest significance level ($p < 0.001$). The critical role of relationship management in the process of adopting new practices, also for the specific case of C2C, has already been underlined by previous research as well as in the qualitative study[72]. This aspect is particularly relevant in the context of C2C certification since the certification process was organized at the time of the survey by EPEA and MBDC and only few other assessment bodies[73]. The collaboration includes the transfer of critical knowledge on the requirements for a C2C certification as well as establishing an intermediate role between companies and suppliers, who sign non-disclosure agreements and reveal recipes or material compositions to the assessment body only. These are key steps on the way to a certification and the tasks can only be undertaken by accredited assessors, such as EPEA and MBDC, so that the companies are somewhat dependent on the collaboration. Thus, it doesn't seem surprising that a good relationship with such a crucial partner positively influences the perceived satisfac-

---

71 P3: *The company's motivations, which are most often expressed in terms of respective expectations, have an influence on the company's satisfaction with the C2C implementation and certification process.* See chapter 3.3 for the detailed elaboration.
72 See chapters 3.2 and 4.1.1.
73 As further elaborated in chapter 2.3.3, the responsibilities of EPEA and MBDC have changed since the survey has been conducted. Both still present accredited assessment bodies, while the C2C Products Innovation Institute takes the role of administering the certification program, e. g. through further developments and updates of the certification criteria Cradle to Cradle Products Innovation Institute (2020).

tion with the overall C2C implementation and certification process. Further studies on C2C certification confirm this observation, e.g. the case study on Herman Miller, which emphasized the high benefits of expertise brought in by MBCD and their important role for on-boarding and collaborating with suppliers with regard to opening up material compositions and proprietary data (Rossi et al., 2006). The role of the assessment bodies appears specifically relevant in the context of sustainability-related innovations, which often show a high complexity and hence require very specific judgement and knowledge (Ketata et al., 2015; Seebode et al., 2012). However, due to the indispensable importance for the certification process, the role of EPEA and similar assessment bodies is also critically discussed, for instance because they don't freely reveal chemical data on C2C compliant materials and composites, so that companies can hardly bypass the collaboration (Toxopeus et al., 2015). This means that in the certification context, a poor relationship might also quickly lead to discontent. In a more extreme case, negative experiences with the certification partner could even result in the complete abandonment of the certification effort, as indicated by a few companies in the qualitative study as well. Such an additional complexity level of undergoing a certification process is particularly relevant for the C2C concept and other voluntary sustainability standards. This was also underlined through comments in the interviews and the survey, when companies reported that they abandoned the C2C certificate, however proceeded with product innovation according to C2C guidelines.

One further explanation for the high effect size can stem from the potential intermediary role of the standard assessors for the company and its customers. Especially when C2C-related activities are complex and not necessarily clear or visible to the end customer, the collaboration with a standard-related partner can offer more transparency and credibility with the help of the provided label (Schons and Steinmeier, 2016).

C2C anchorage: The depth and breadth of implementation, captured in the factor 'C2C anchorage', varies between companies adopting C2C and also impacts their retrospective on the C2C implementation process, reflected by a slightly negative effect revealed in the regression analysis ($p < 0.1$). One potential explanation could be based on the system-level perspective of C2C, if implemented substantively. As it already became evident, C2C implementation touches many areas and affects various groups of stakeholders, which might complicate a thorough and extensive implementation process. For some C2C innovations, the product development per se is not enough so that additional effort is required in order to change the business model, adapt standard operating procedures or customer interaction routines. Hence, besides the aforementioned knowledge and relationship management, the implementation might also call for the adaptation of the company's business model (Boons and Lüdeke-Freund,

2013; Diaz Lopez et al., 2018; Nidumolu et al., 2009). This in turn, might require the onboarding of additional internal and external stakeholders who were not related to the innovation process so far or who are used to different processes and work routines (Boons and Lüdeke-Freund, 2013; Diaz Lopez et al., 2018; Ghisellini et al., 2016). If not only the product development undergoes a change process but also the business model needs adaptation accordingly, overcoming such established habits within the company, which Boons and Lüdeke-Freund (2013) refer to as "institutionalized organizational memory", might be particularly challenging.

Revising organizational habits can also potentially urge companies to prove a degree of "willingness to cannibalize", meaning that a company would be prepared and willing to down-prioritize existing products or technologies in favour of new investments, e.g. in radical innovations (Chandy and Tellis, 1998; Danneels, 2008). However, the difference between such extensive implementation activities and more symbolic actions, also known as 'greenwashing' in the sustainability research field (e.g. Delmas and Burbano, 2011), is not always visible to the customer (Petersen and Brockhaus, 2017). Hence, one potential explanation for the negative effect of C2C anchorage could also lie in a deceptive outcome of C2C implementation efforts, depending on the group of stakeholders which is addressed by these efforts (Schons and Steinmeier, 2016). In the concrete context of C2C innovations, this might be further explained by the certification scheme which involves financial resources and a mandatory commitment to defined requirements in order to achieve a certain certification standard in the first place, but also in the subsequent re-certification (see chapter 2.3.3). Engaging in such a certification scheme, including the related expenses, might make it more difficult for companies to decouple actual practices from more symbolic actions (Christmann and Taylor, 2006).

Not exclusively linked to sustainability or CE-related innovations but to innovative capabilities per se, the concept of 'absorptive capacity' has been the subject of various studies in the innovation management research area already (see Zahra and George, 2002 for a detailed literature review). Coined by Cohen and Levinthal (1990) decades ago, this term describes the "ability of a firm to recognize the value of new, external information, assimilate it, and apply it to commercial ends" (Cohen and Levinthal, 1990, p. 128). In the context of C2C innovations with its aforementioned characteristics, there is certainly a difference between companies with a greater or smaller need to 'unlearn' organizational routines and respectively their ability to succeed in this process (Cohen and Levinthal, 1990; Ketata et al., 2015; Seebode et al., 2012; Schmitt and Hansen, 2018).

Overall, the factor C2C anchorage reflects the system-level perspective (Niero et al., 2017) needed during C2C implementation. This complexity and the manifold interlinks between numerous stakeholders often imply high efforts throughout the company, up- and downstream, which in turn can explain the negative effect of 'C2C anchorage' on the company satisfaction.

Integration into supply and value chain: Besides the relationship and knowledge management, the factor 'integration into supply and value chain' captured the company's efforts to include various actors and stakeholders in the C2C implementation process. Since circular innovations can require a complete re-design of the product and product development depends on the sourcing of new materials, the early involvement of suppliers in the search process for compliant substances proves particularly critical. The re-arrangement of components or introduction of new technologies and components has been discussed in the research field of architectural and modular innovations as well (Abernathy and Clark, 1985; Bozdogan et al., 1998; Henderson and Clark, 1990). Depending on the type of innovation that the C2C adoption presents for the respective company, the implementation process implies different approaches with regard to organizing knowledge management for a company.

This also points at the important role of the various value and supply chain actors, which need to be accurately managed and interlinked, also including an open and trustful communication between the various stakeholders (Schmitt and Hansen, 2017). Companies could regard their supply base more as knowledge partners than just material providers and at the same time enforce their exchange with internal employees, e.g. with R&D teams or engineers (Bozdogan et al., 1998; Rossi et al., 2006). It also suggests the need for a holistic approach, so that not only direct customers or direct suppliers are involved, but also taking into account various players and stakeholders further up or down the value and supply chain (Diaz Lopez et al., 2018; Niero et al., 2017; Paramanathan et al., 2004), which hints at relevant capabilities that potentially need to be adapted in the CE context (Vanpoucke et al., 2014). The relatively high effect size and significance level ($p < 0.01$) of the factor underline the necessity of such an active management of the internal and external actors along the supply and value chain to achieve a satisfactory outcome of the implementation and certification process[74].

Further, the critical role of supply chain management in the particular context of sustainable innovations and its importance for CE efforts has already been pointed out by diverse scholars. For instance, the idea of a closed-loop supply chain describes the process of "taking back products from customers and recovering added value by reusing the entire product, and/or some of its modules, components, and parts"

---

74 The recent work by Hansen and Schmitt (2020) is one of first to explicitly research how C2C implementation affects the value chain with the help of a longitudinal case study. The authors identify different barriers and potential mechanisms to overcome these and put particular emphasis on innovation communities and collaboration.

Discussion of findings

(Guide and Van Wassenhove, 2009, p. 10) which leads to a maximized value creation considering the entire life cycle of a product. Thus, when transferring the idea of circularity into practice, this process can play a major role since it involves the critical points that are affected by a C2C implementation such as sourcing, product design or reverse logistics (Carrillo-Hermosilla et al., 2010; De Pauw et al., 2013; Morana and Seuring, 2007; Kalogerakis et al., 2015). Hereby, the envisaged value creation can be supported by an appropriate adaptation of the value and supply chain (Jayaraman and Luo, 2007).

Degree of change: The underlying items of the factor 'degree of change' reflected the need for fundamental changes to existing product(s) and processes as well as for the adaptation of the innovation process. However, the factor didn't yield sufficiently high significance levels in the regression analysis. A statistical examination of potential reasons couldn't provide further insights, e. g. comparing the variance of the factors. Further, no apparent differences were identified in the data with respect to companies holding the C2C certificate for a longer or shorter period of time which might have provided an indication of a time-lag effect. From a contextual lens, the absence of an effect could be explained with the potentially secondary role of the existing products applying for the certificate. As some companies mentioned during the interview series, there was no need for change since all criteria of the C2C certification scheme were already fulfilled. However, other companies were forced to undergo a larger change process in order to achieve certification. Yet, the degree of required change might not play a crucial role per se but rather the company's ability to succeed in such a process. Since the change management process depends on various organizational and human resource aspects (Kotter, 1995), the factor might not reflect the actual root cause for a satisfactory outcome of C2C implementation. One other potential explanation could be the position of the respective company on the technological S-curve (Christensen, 1992; Foster, 1986). This would indicate that the key product attributes, which are critical for the purchasing decision of the customer, might be affected in different ways by a necessary change process and may also be more or less visible to the customer (Foster, 1986).

Company size: Reflecting the organizational complexity, operationalized through the number of employees (chapter 4.1.1), the factor 'company size' was checked for its moderating role in the organizational context. Even though not necessarily linked to sustainability, previous studies have already researched the influence of company size on the adoption behaviour, e. g. an early study of Hannan and McDowell (1984) on technology adoption with regard to organizational size and industry concentration. Yet, it can be corroborated that instead of looking at organizational size as a standalone factor, it is more accurate to understand "under what conditions particular aspects of size are important for what other organizational characteristics" (Kimberly, 1976, p. 586). The absence of a direct effect on the company's satisfaction in the regression analysis underlines this argument and allows for the assumption that the company size per se doesn't influence how companies handle C2C implementation and certification. This also implies that the C2C concept has a broad suitability across all company sizes. However, the significant positive interaction effect ($p < 0.05$) of company size and the factor 'integration into supply and value chain' points out that the larger the company, the more important it becomes to integrate and align all actors of supply and value chain. Given the assumption that proper management of a closed-loop supply chain already involves various actors and demands a complex relationship management (Östlin et al., 2008), this complexity would probably further increase assuming that organizational complexity often increases with the company size.

Having integrated the findings with regard to the developed propositions, the degree of support for P4, addressing the effects of organizational context factors can be rated as contradictory. On the one hand, the criticality of internal and external stakeholders (proposition P4a[75]) could be supported by significant effects in the regression analysis. The influence of necessary adaptations of products and processes for a successful outcome, expressed in proposition P4b[76], on the other hand, couldn't be proved valid in the analysis.

Proposition P5[77] can be supported with a stronger degree since an increased organizational complexity influences how various actors across the supply and value chain need to be integrated.

New product success: The recurring theme of the market response on the developed C2C innovations was captured in proposition P6[78]. As shown in the regression analysis, the factor 'new product success' was the only covariate with a highly positive

---

75 P4a: *The adoption of the C2C certification is influenced by the organizational context in such way that internal and external stakeholders are critical for a successful implementation.* See chapter 3.3 for detailed elaboration.

76 P4b: *Current processes as well as existing skills and capabilities need to be reviewed and potentially adapted in order to achieve satisfactory results* See chapter 3.3 for detailed elaboration.

77 P5: *The effect of organizational context factors on the company's satisfaction with C2C implementation and certification also depends on the prevailing organizational complexity which influences C2C adoption within the company.* See chapter 3.3 for detailed elaboration.

78 P6: *In contrast to a more equivocal relationship between motivational and organizational factors with the company satisfaction, the market success of the C2C certified product(s) clearly positively influences the eventual satisfaction with C2C implementation and certification.* See chapter 3.3 for detailed elaboration.

and significant (p < 0.001) effect on the company's satisfaction with C2C, thus providing strong support for proposition P6. The somehow expected positive impact also points at the need to better visualize intangible benefits of CE efforts that can't be measured directly with regard to financial returns.

In the case of sustainability-related innovations, the customer role can be particularly complex since the shift towards a Circular Economy also requires a shift of current consumer patterns (chapter 2.2). This complicates the role of companies as they need to take into account that customers might not always appreciate innovations, which result from an adopted innovation process based on new CE-related criteria (Petersen and Brockhaus, 2017). Furthermore, as was pointed out in the interviews (chapter 3.2), when companies increase pricing to compensate for the additional development efforts, this might even lead to a smaller customer base and thus increase the discontent with the respective sustainability effort, such as the C2C certification. The qualitative study also revealed that the customer perception can vary depending on the industry sector and the affinity of the customers for sustainability, e. g. in some cases companies were even pro-actively addressed by the customers who were looking for C2C certified products. Thus, an accurate analysis of the customer base and current market developments is not only particularly necessary but also pushes the boundaries of different functions and departments in a company. On the one hand, once customer needs have been identified, they need to be aligned with the goals of the sustainability-related practice and integrated into the product development process as early as possible (Abele et al., 2005; Carrillo-Hermosilla et al., 2010). On the other hand, this further underlines the strong design focus of CE-related innovations (chapter 2.2 and 2.3) and emphasizes how they are intertwined with the fuzzy front end of innovation[79] (Bocken et al., 2014). Thus, incorporating CE principles in the earliest phases of the design process, including a clearer orientation towards customer needs, can further contribute to a more satisfactory implementation process.

The predevelopment activities in the innovation process have been postulated as a success factor by numerous researchers already (see Koen et al. (2002) for a detailed review) and a differentiating factor between successes and failures of new products (Cooper, 1988). The focus on the earliest design phases can also contribute to minimizing cost of change, in particular when new concepts or practices are introduced (Abele et al., 2005; Rossi et al., 2006). The fuzzy front end of innovation also needs regular revisions depending on developments in the outside world or emerging science and technology (Koen et al., 2002) and herewith offers potential for the integration of CE-related factors. However, this might also lead to longer development times, especially for companies which engage with sustainability-related innovations

---

[79] See Verworn (2004) for a definition and detailed analysis of the fuzzy front end in product development.

for the first time (Rossi et al., 2006). Thus, the concentration on the early phases of the product development proves particularly promising to achieve product success and hence a satisfactory outcome after undergoing the implementation or certification process. This could encourage companies to hold on to the sustainability-related efforts and henceforth foster the long-term establishment of CE principles in business.

Ultimately, focusing on specific customer needs under consideration of sustainability-driven goals can also force companies to revise "current ways of fulfilling customer needs" (Hansen et al., 2009, p. 693), which can also enhance innovativeness with respect to prevalent business models (chapter 6.3).

5.2 Building an organizational environment to foster CE innovations

In order to answer how CE standards can successfully be anchored at the company-level in the long term (RQ3), the results of the qualitative study (chapter 3.2) and the quantitative exploration (chapter 4.5.3) shall be further discussed and the related proposition P1, reflecting the overarching process of balancing benefits and costs[80] will be validated. The discussion also forms the basis for the following implications for managerial practice (chapter 6.2).

As pointed out in the previous section, the analysis underlined the need for a thorough relationship management with the certification partner, but also with actors along the supply and value chain, including internal and external stakeholders and key decision-makers. This emphasizes a systemic view when it comes to C2C (chapter 2.2.1), and affects different areas on the organizational level.

First, the necessity to take on a holistic perspective can pose major challenges for internal collaborations between different departments, such as R&D, supply chain management or marketing and sales since the departments might not always be oriented towards the same objectives. Particularly, this applies for financially driven functions such as accounting, which were often found to be in a counter-position to the C2C initiating units. As in most companies the prevailing rationale for business decisions is based on the (short-term) financial viability of new products or practices, companies would require new evaluation standards of sustainability-related innovations, such as C2C. This could provide decision-makers with a new rationale

---

[80] P1: *A company's decision whether to further pursue C2C depends on the satisfaction with the outcomes. In turn, the satisfaction is influenced by a company-specific and differentiated balancing of various factors. These factors include initial motives to engage with C2C, the organizational processes (enablers or barriers) involved, and the final customer feedback. See chapter 3.3 for detailed elaboration.*

for their choices, for example by adapting incentive systems (Chappin et al., 2015; Schaltegger and Hörisch, 2017). With respect to the critical role of the market perception of the C2C innovation (captured in the factor 'new product success') and its relevance for the eventual satisfaction (chapter 5.1.2), the development of a more differentiated evaluation system for circular innovations appears even more acute. Additionally, the coordination efforts need not only to be modified internally but also with regard to external partners, particularly with suppliers and their assessment (Bozdogan et al., 1998).

The challenges of a transition process and the related integration trade-offs, for example with regard to exploration versus exploitation efforts, has been amply described in the work of Hansen et al. (2018) on structural ambidexterity. The authors provide a framework which facilitates an accurate analysis of the available mechanisms for a successful organizational design depending on the respective stage in the transition process. Though not exclusively linked to circular innovations, the work highlights the high complexity of the required trade-offs for a company that sets out to integrate new technologies. As described by some interviewees during the qualitative study, the internal controversies led to frustration or even abandoning the implementation effort (see chapter 3.2). Hence, making the necessary choices visible and creating awareness for these could present an important step during for the implementation success.

Further, the role of top management and critical decision-makers has already been identified as an important factor for persevering in the C2C efforts and overcoming internal and external impediments (chapter 5.1.1). The strategic fit of the selected CE practice further enhances a consistent implementation path and allows for consequently incorporating new standards into the innovation process. This demands from the respective decision-makers to shape a developmental instead of a hierarchical culture which stands for a flexible orientation of the company, also including external developments (Büschgens et al., 2013).

Second, when looking at the management of the innovation process, the integration of different views and their incorporation into the new product development can trigger new practices that apply particularly well for the context of circular innovations. Here, the open innovation paradigm (Chesbrough, 2003) could present such a suitable approach. In contrast to the closed innovation paradigm, which builds on the idea of managing the entire innovation process from idea generation to production and distribution with in-house knowledge sources and resources, in the open innovation paradigm "valuable ideas can come from inside or outside the company and can go to market from inside or outside the company as well" (Chesbrough, 2003, p. 43). In this context, the critical role of the certification-related organizations during the C2C implementation process (chapter 4.5.3) already emphasized that many companies in the transition phase struggle to generate or apply new knowledge that conforms to circular standards. In these cases, opening up the innovation process could be a valuable source of knowledge and spin-offs or open source efforts could leverage circular activities. In the particular case of C2C certification, the qualitative study revealed that some companies needed to invest more than others in researching substitutes for toxic or harmful substances in order to receive the certification and would have benefited from increased knowledge diffusion, e. g. from fellow competitors who have already completed the certification (chapter 3.2). However, so far, the certification process is accompanied by non-disclosure agreements with the related organizations (Toxopeus et al., 2015).

To foster the systemic perspective and better identify the sources of value creation, the *Value Chain Framework* (Figure 27) can present a suitable approach allowing for an accurate assessment of positive and negative impacts of the company's core business processes, particularly in the sustainability context (Porter, 1985; Porter and Kramer, 2006). Originally applied to the social impact of the value chain, Porter and Kramer (2006) developed the overview of the "activities a company engages in while doing business [...]" in order to "[...] identify the positive and negative social impact of those activities" (Porter and Kramer, 2006, p. 8).

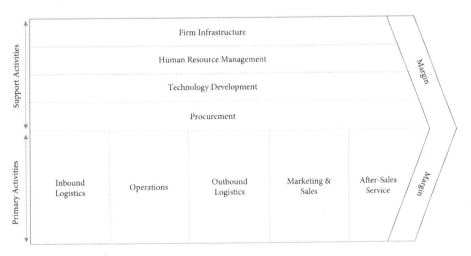

Figure 27: Value Chain framework[81]

When applied to the context of the transition towards a Circular Economy, such a perspective could leverage the potential for value creation through engaging in appropriate CE-related practices by more effectively involving all necessary partners and considering critical process stages.

---

81   Source: Porter and Kramer (2006).

For example, the activities falling under 'Technology Development' could be adapted in a way that they better conform to sustainability standards, such as in the C2C case. This would probably require new sources for knowledge collection and at the same time promote new collaboration forms (Porter and Kramer, 2006). The introduction of new signals to the company's search strategies could also facilitate the recognition and potentially faster adaptation of new technologies (Seebode et al., 2012). Therefore, a company could be better enabled to adopt CE-related practices without disregarding the value creation potential for its business (Porter and Kramer, 2006; Porter and Van der Linde, 1995).

Reflecting the importance of a strategic fit together with established routines, the ReSOLVE framework, developed by the Ellen MacArthur Foundation, can further enhance C2C implementation by providing a new basis for decision-making for the operational level. To facilitate the embrace of a new perspective with respect to circularity of innovations, the framework helps to translate the CE idea into six business actions. The name is the acronym for the six steps being Regenerate, Share, Optimise, Loop, Virtualise, and Exchange (Ellen MacArthur Foundation, 2015a, 2015b).

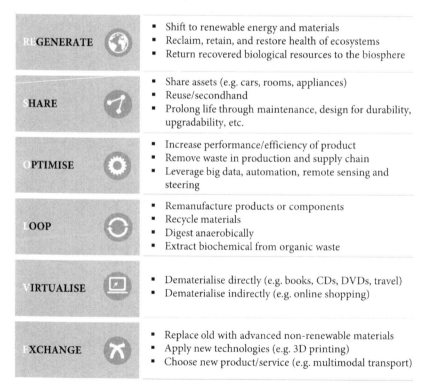

Figure 28: The ReSOLVE framework[82]

---

82   Source: Ellen MacArthur Foundation (2015b).

Figure 28 presents the relevant steps defined for every business action, which resulted from conceptual work by the foundation as well as in collaboration with pioneering companies who already engage in CE. Ultimately, the proposition P1 allows for a strong degree of support regarding the complex balance of implementation motivations, the internal and external organizational environment, and customer-related considerations. It appears that companies don't necessarily stick to the chosen path when it comes to implementing sustainability-related practices or striving for certification, as in the case of C2C. The engagement seems to be more of a process in which the costs and benefits are evaluated on a continuous basis, considering efforts and actual returns compared to expected ones. In such way, companies can also decide to abandon the initiation of circular innovations, the certification application, or both, depending on the level of their discontent.

Thus, a framework or tool, such as the presented value chain framework, which helps to more objectively balance the different perspectives would present a helpful decision guidance. As described earlier, one important factor in the evaluation is also the presence of routines and habits that a company would first need to 'unlearn' before it can further adopt new circular concepts (Becker, 2008; Seebode et al., 2012; Van de Ven and Polley, 1992). This would imply that all involved departments, also including corporate communications, procurement and sales teams, need to be made familiar with a new measurement standard of value creation and capture, which doesn't necessarily correspond to the prevalent ways of measuring financial success or legitimacy (Martín-Peña et al., 2014; Schaltegger and Hörisch, 2017).

Here, a more granular guidance could support the development of decision criteria on the organizational level. This could be achieved with by defining company-specific action steps according to the ReSOLVE framework that should be promoted internally and externally during the implementation process. Especially, when trade-offs need to be made, such guidelines can leverage the collaboration between departments by pointing out a common direction when integrating different actors. Thus, the framework can be an appropriate means of providing points of references for a customized circular pathway for companies[83]. For companies requiring more complex transition or integration efforts, the management of trade-offs would become even more important and needs to be thoroughly managed with regard to structural ambidexterity (Hansen et al., 2018).

---

83   In a more recent work by Mendoza et al. (2017), the ReSOLVE framework was extended with a special focus on implementation after reviewing multitudinous tools and frameworks and their suitability with ReSOLVE criteria. The work consists of a comprehensive review of research frameworks in the area of Circular Economy and concludes that most suggestions are still deficient with respect to providing guidelines at the implementation level (Mendoza et al., 2017).

Discussion of findings

Conclusively, though companies need to balance many different objectives and sometimes adhere to contradictory expectations by various stakeholders, the consideration of the critical organizational context factors and active arrangement of a respective corporate environment can facilitate a successful adoption of sustainability-related practices, such as the C2C certification scheme, and ultimately leverage a long-term shift to a Circular Economy.

6 Conclusion and implications

The aim of the present thesis was to cast light on the initiation and implementation of CE-related innovations at the organizational level and embed the CE phenomenon in the innovation management context. Since research on a clear-cut meaning of sustainable innovation is still somehow disjointed, the Cradle to Cradle standard was applied as a case example and thus defined a concrete empirical field, from which further conclusions can be drawn. The results of the qualitative and quantitative exploration contribute to existing theory (chapter 6.1) and also provide implications for practice (chapter 6.2). In chapter 6.3, the limitations of the present work are elaborated and future research opportunities derived.

6.1 Implications for theory

The idea of sustainability has been widely researched under different angles and in various contexts for decades. The concept of Circular Economy presents additional facets to the discussion of sustainable innovations and marks the way to a paradigm shift. The present work contributes to the research field of Circular Economy in three major ways.

First, to the best of the author's knowledge, this study is the first to quantitatively explore the latent variables that influence the shift towards Circular Economy, operationalized through the C2C concept in order to reflect the organizational focus. Hereby, the analysis enriches the understanding of motivational factors and organizational enablers in a research field that so far has been dominated by analyses at an anecdotal level. As concluded by Lieder and Rashid (2016), the CE research landscape is hitherto mainly fragmented and lacks evidence at the implementation level. These shortcomings however might hamper the transition towards CE since the motives and necessary activities required to succeed in the CE shift are scarce and lack evidence, particularly with regard to potential benefits for the companies (Lieder and Rashid, 2016). This endorses the classification of CE as an umbrella term in its validity challenge phase and the need for further work, particularly on concrete CE strategies that answer "how" questions with respect to CE implementation (Blomsma and Brennan, 2017). The methodological approach and organizational focus of the present work hence address this research need and deepen the understanding of why and how companies manage the implementation of a CE-related concept.

Second, as a basis for the statistical analysis, variables from related research fields are combined with newly developed items resulting from the qualitative data analysis by statistically aggregating them to factors (chapter 4.4). Thus, the study provides a set of constructs to operationalize motivations for initiating CE-related activities as well as organizational enablers that can leverage a satisfying outcome. This contributes to the hitherto rather scarce empirical basis on the role of initial motivations of companies which engage in CE. The results corroborate findings on the existence of divergent motives, such as profit- or legitimacy seeking motives (Bansal and Roth, 2000; Schaltegger and Hörisch, 2017) and further add the perspective of a fit between the selected actions, here pursuing C2C certification, with the company strategy. This presents a research area that has received little attention so far. Therefore, the results can provide additional insights on symbolic versus substantial implementation efforts (Delmas and Burbano, 2011; Schons and Steinmeier, 2016). The study suggests that more satisfactory implementation results can be achieved when companies opt for a concept that is accurately aligned with the company's strategy and philosophy.

As part of the factor 'strategic fit', management support also proves critical and advises the involvement of decision-makers when engaging in CE-related concepts. The present analysis empirically supports the decisive role of top management for a successful implementation and certification process, which has been discussed almost in unison in previous research already, on the one hand for new product development more in general (e.g. Cooper, 1988) and on the other hand in the context of sustainable innovations (e.g. Abele et al., 2005).

Since the C2C certification applied as empirical field and responding companies had a common understanding of the queried actions, the study results are particularly insightful. So far, a major part of previous studies includes a manifoldness of activities leaving more room for subjective interpretations on the actual reasons to engage in sustainability-related actions (Paramanathan et al., 2004). Further, there has been limited research on the potential impact of initial motivations on the diffusion of CE. Thus, by applying a quantitative method, the present work enriches the understanding on how motives affect the implementation success, expressed through company satisfaction.

Ultimately, the results provide empirical support for the development of a CE stimulating organizational environment by pointing out barriers and enablers of implementation. Previous research already underlined the importance of internal and external contextual factors as well as the relevance of the activities for a company (Abele et al., 2005; Bansal and Roth, 2000). The results successfully expand on this by identifying the impact of organizational factors which hinder or foster the implementation success. By pointing at the determinants which drive company satisfaction with C2C implementation, this work can make a contribution to accelerating the diffusion of CE-related concepts. Even though extant theory on success factors of new product development and innovation management (e.g. Cooper and Kleinschmidt, 1991; Hart, 1993; Koen et al., 2002) provides critical guidelines that also apply for sustainability-oriented innovations, such as C2C, the present work casts light on additional layers of complexity (Hall and Vredenburg, 2003) related to the integration of the CE concept into innovation management.

6.2 Implications for practice

As one of the first empirical studies to investigate a large set of C2C adopting companies, the present work provides implications for managerial practice with respect to the initiation and implementation of a CE-related concept, such as C2C.

Drivers for implementation

Besides more researched motives such as legitimacy- or profit-seeking, the present research emphasizes a third motivational factor which is based on a strategic matching of the selected circular concept with the company's strategy. This points at the importance of an accurate selection of the concrete CE concept to be implemented. Following competitors or industry trends in a one-size-fits-all manner would probably lead to a less satisfying implementation. As in the present case of C2C, the strategic fit of the concept with the company's values and targets proved critical for a positive overall assessment after its implementation. In particular, when financial resources are required, not only for the implementation itself but also, for instance, for a related certification program, such as in the case of C2C, decision-makers who ensure the suitability of the CE concept with the prevailing corporate strategy might more likely achieve satisfactory outcomes. However, traditional cost-benefit considerations often fail to properly include sustainability aspects, most probably due to the deficient measurability of the outcome, which has also been elaborated by some interview partners (see chapter 3.2). This calls for innovative methods which are more suitable for the CE context and can help companies to identify the most appropriate set of activities based on the respective company setting, e.g. the sustainability innovation cube as suggested by Hansen et al. (2009). Furthermore, the relevance of a strategic fit draws attention to the importance of employee acceptance for the adoption of sustainability-related concepts in order to succeed in a change management process (Kotter, 1995; Seebode et al., 2012).

The negative effect of the factor 'competitive pressure' on the implementation satisfaction (see chapter 4.5) underlines the complexity of organizational responses to external pressure (Crilly et al., 2012). This effect can be attributed to various causes, which are sometimes also subject to company-specific factors (chapter 5.1.1). To avoid negative effects resulting from the reactive character of their effort, companies could more actively take risk-reducing measures. For instance, the articulated difficulty of knowledge building for companies which showed a higher need for product and process adaptations (see chapter 3.2) could be mitigated by the introduction of a transdisciplinary collaboration (Schaltegger et al., 2013) and an active management of trade-off mechanisms (see chapter 5.2) during the critical phases of the integration process (Hansen et al., 2018). As suggested by the concept of Open Innovation (Chesbrough, 2003), this would imply not only cross-functionality at the company-level but also academia-practice collaboration, including NGOs, customers or other institutions. Opening up technology development and R&D (chapter 5.2) in the sense of open innovation could thus facilitate earlier detection of important trends and technologies and their implementation, thus supporting companies to move from a reactive to a more pro-active approach towards circularity (Schaltegger et al., 2013).

Organizational context factors
The analysis provided evidence on the importance of an active and accurate coordination of internal and external partners. The need for a thorough coordination was reflected in the factor on the relationship with the certification partner as well as on the integration of supply and value chain. The practical implications build on the discussed results in three major ways.

First, as elaborated in chapter 5.2, frameworks that focus on the holistic system view of managing the shift towards CE can provide useful means in order to align the critical stakeholders involved. The application of the value chain framework and the ReSOLVE framework could thus enhance the onboarding process of critical partners, not only within but also outside the company. Thus, the objectives of internal functions could be better aligned and the communication with external parties, such as suppliers, facilitated. Referring to the results of the moderation analysis (chapter 4.5.3.3), larger companies are urged to pursue the integration into supply and value chain even more. Especially in cases when companies don't have clear visibility into the processes of their supply chain, e.g. due to restricted transparency with regard to sources and manufacturing details of their supply base, the CE engagement can reveal such deficits. Besides, the supply chain plays an important role for realization of circular innovations since reverse logistics are an essential element when it comes to closing the loop in a circular way (Carrillo-Hermosilla et al., 2010; Morana and Seuring, 2007). However, the onboarding of all actors involved up or down the value chain is important in order to achieve a satisfactory outcome. A close collaboration with a more 'neutral' third party could facilitate this effort. In the case of C2C certification, this role was most often assigned to EPEA or MBDC, also considering their network which was often needed to foster knowledge creation and transfer with experts. Further, the assessment bodies related to the certification standard can contribute by directly negotiating with suppliers and obtaining critical information, e.g. through signing non-disclosure agreements. When companies don't engage in a specific certification program, CE-related efforts could then be supported by other alliances, e.g. with academia or other expert groups.

Second, the open innovation paradigm also provides the opportunity to more actively and successfully manage the various stakeholders of a company in its CE engagement (Seebode et al., 2012). For example, the integration of customers into the idea generation process promises to be highly beneficial, also given the empirical results on the essential role of the market success of circular innovations for the company satisfaction. In the case of C2C adoption, including the customer perspective could provide valuable insights with regard to the crucial product characteristics, no matter whether in the B2B sector or addressing the private end consumer. This would encourage the focus on required functionalities and might help companies during the adaptation of products conforming to C2C. Although this would demand a revision of the extant innovation process for some companies more than for others, the results underline the importance of integrating the CE standards as early as possible in the product development process (Bocken et al., 2014). Further, since the collection of rather new and complex knowledge has been emphasized as a critical part throughout the research studies, an open innovation approach could leverage this process.

Third, depending on the necessary adaptation of the corporate culture and internal processes, adopting a CE concept or pursuing a respective certification should be accompanied by a thorough change management process for many companies. In some cases, a substantial anchoring of the new circular principles might necessitate radical changes of prevailing processes or routines and the need for resources might also raise expectations for success. Therefore, the top management or critical decision makers are advised to 'walk the talk' to empower company employees and support the change process including overcoming potential initial hurdles (Kotter, 1995).

The role of government regulation and policies

Beyond the implications for managerial practice, a leveraging role could also be related to the influence of legislation and political incentives in the CE context. As the research results indicated, the process of implementing CE-related concepts is complex and in many cases requires substantial effort. At the same time, the success of the new products significantly impacts the satisfaction with the implementation and certification process. Yet, market success doesn't necessarily have to be attributed solely to the customer reception. The financial viability of the developed circular innovations can also be influenced by political incentives or regulations. Still, the role of regulation and policies seems somehow debatable.

Even though frequently discussed in theory, a comprehensive review of potential barriers to CE of Kirchherr et al. (2018) identified the group of regulatory barriers ranking in the midfield compared to cultural or market barriers. However, assuming the interaction effects between different barrier groups, the role of regulations and policies proves essential. For instance, when considering market barriers such as lower virgin material prices, disadvantageous taxation systems or the high up-front investments as frequently mentioned during the interview series (chapter 3.2), CE-enabling legislation could positively impact the transition towards CE and advancing the development of circular innovations (Horbach et al., 2012; Kirchherr et al., 2018). Even though different initiatives have been launched to foster the transition towards CE, such as the Circular Economy Action Plan of the EU commission (chapter 1.1), the current legislative conditions still seem to leave room for action. Strategies could vary from more pragmatic interventions like a preferential treatment of circular products in public procurement to more radical approaches like abandoning current subsidy programmes that might counteract the application of new materials or development of circular business models (Del Río et al., 2010; Kirchherr et al., 2018). The development and adoption of new technologies currently seems to be trapped in a sort of "vicious circle" since "they [emerging technologies] are not adopted because they are too expensive, and they are too expensive because they are not adopted" (Del Río et al., 2010, p. 543). Targeted policies and governmental strategies, like CE funding programmes or specific tax incentives for companies selling CE products could surely present an important element in breaking this logic and thus accelerating the transition towards CE.

6.3 Limitations and avenues for future research

The present study intended to cast further light on why and how companies engage in CE practices. Given the explorative character of the work, the C2C certification has been selected as an empirical field in order to ensure comparability due to clear certification criteria and a common understanding among survey respondents. Still, certain limitations arise which should be considered for an accurate interpretation of the results at hand.

First, the absence of a measurable definition of CE prompted numerous academic discussions on the taxonomy as well as similarities and differences between the schools of thought related to sustainability, such as eco-innovations or sustainable innovations, industrial ecology or circular economy (chapter 2.1 and 2.2). To be able to build on a suitable empirical basis, the present analysis was conducted in the context of the C2C certification program. While the selection of this empirical field ensured the comparability of survey results and allowed one of the first quantitative explorations in this research field, the generalizability might be limited due to characteristics specific to the C2C certification.

Second, the survey design and involvement of one key informant per company (chapter 4.1.2) might be prone to biasing effects such as the common method bias or informant bias. Preventive measures in the survey design and execution were undertaken in order to minimize this risk. These were complemented by statistical remedies prior to the regression analysis (chapter 4.5.2). Further, to reduce potential success biasing (Diaz Lopez et al., 2018), the survey consisted of various constructs intending to draw a comprehensive picture of implementation motives and processes, by including questions on success and failure experiences and explicitly targeting companies which had already abandoned the C2C certificate at the time of the survey. Altogether, the thorough assessment of conducted preventive measures and statistical tests indicate the absence of these bias effects.

Third, given the lack of empirical evidence in the CE research field with respect to implementation at the organizational level, the present analysis aimed at answering the 'why' and 'how' questions formulated in the research questions by performing a quantitative exploration (Bamberger and Ang, 2016). Accounting for the deficit of constructs that have been tested in this context already, the study built on related research fields and has been complemented by insights from the precedent qualitative study. The use of propositions instead of hypotheses also reflected the absence of clear indications on the expected direction of effects in most of the cases. This might expose the study to a limited representation of relevant factors. However, given the robust results of the statistical analysis, such as the relatively high value of the explained variance (chapter 4.5.3), it appears that the developed research model is adequately suitable for the selected research context. Yet, a re-assessment of the developed factors by means of a confirmatory analysis would strengthen the interpretation of different motivational and organizational factors and uncover their role for a company's overall implementation satisfaction.

In the presence of the mentioned limitations, the present study can also provide fertile ground for further analysis in multiple ways, on the one hand specific to the present work, on the other hand concerning the research field more in general.

Building on the present research endeavour, two main anchors for further analysis can be derived. Conducted with a different certification standard, which is also closely related to the CE concept, a comparable analysis could facilitate the comparison of results and the derivation of more general insights. The results of the qualitative and quantitative study could further provide the basis for hypothesis development and a confirmatory investigation in a similar setting. In addition, based on the assumption that a satisfactory C2C implementation leads to a stronger CE commitment of companies, a longitudinal perspective would provide a valuable addition to the present results by investigating whether companies actually prolonged or extended their certification activities.

After having classified CE as an umbrella term, currently in the 'validity challenge' phase (chapter 2.2.), the concept appears to be in transition to the 'further work' stage (Blomsma and Brennan, 2017). This opens up multifaceted research possibilities, especially at the organizational level as pointed out by the present results.

The first area, in which further research would contribute to a valuable enhancement of the topic, would be the call for transdisciplinarity (Schaltegger et al., 2013) during the CE implementation process and its relevance for a lasting paradigm shift. Given the complexity of integrating different stakeholders in- and outside the company, including external knowledge sources, the adoption of CE standards poses major challenges for the innovation management process. This area presents a major research opportunity with respect to setting up the right corporate environment and identifying success factors for the communication and collaboration, potentially even specific to certain industries or company sizes. In this context, the company's absorptive capacity (Cohen and Levinthal, 1990) can play a major role for success and deserves further research attention (Schmitt and Hansen, 2018).

Already identified as a critical factor for the realization of circular or sustainable innovation, the management of the early development phases in the fuzzy front end of the innovation process presents an additional complexity layer that merits exploration, particularly taking into account the system perspective of the CE concept (Bocken et al., 2014; Stock et al., 2017). For example, the sourcing and application of new, benign materials appears to be peculiarly challenging, and has so far hardly been addressed, especially when it comes to concrete implementation steps. The idea of an intelligent materials pooling system (Braungart et al., 2007) could enhance the management of the technical cycle, however the elaboration of such systems has so far received only little attention.

Last but not least, a more comprehensive integration of the customer perspective in the CE transition might provide valuable insights on how to foster CE transition (Horbach et al., 2012). It appears that the market seem to be 'not ready' for circular products and circular business models, which even more strongly depend on the customer. As long as the market perception is deemed insufficient by companies, the shift towards a CE will be difficult to realize in the long run. Therefore, research scholars could contribute to overcoming the current consumption patterns not only by investigating the role of product development but potentially also how innovations can educate consumers in order to substantially anchor the CE paradigm in the economic system.

7 References

Abele, E., Anderl, R. and Birkhofer, H. (Eds.) (2005), *Environmentally-Friendly Product Development: Methods and Tools*, Springer, London.

Abernathy, W.J. and Clark, K.B. (1985), "Innovation: Mapping the winds of creative destruction", *Research Policy*, 14: 3–22.

Adams, R., Jeanrenaud, S., Bessant, J., Denyer, D. and Overy, P. (2018), "Sustainability-oriented innovation: A systematic review", *International Journal of Management Reviews*, 18 (2): 180–205.

Armstrong, J.S. and Overton, T.S. (1977), "Estimating Nonresponse Bias in Mail Surveys", *Journal of Marketing Research*, 14 (3): 396–402.

Backhaus, K., Erichson, B., Plinke, W. and Weiber, R. (2016), *Multivariate Analysemethoden: Eine anwendungsorientierte Einführung*, 14th ed., Springer, Berlin, Heidelberg.

Bakker, C.A., Wever, R., Teoh, C. and De Clercq, S. (2010), "Designing cradle-to-cradle products: a reality check", *International Journal of Sustainable Engineering*, 3 (1): 2–8.

Bamberger, P. and Ang, S. (2016), "The Quantitative Discovery. What is it and How to Get it Published", *Academy of Management Discoveries*, 2 (1): 1–6.

Bansal, P. and Clelland, I. (2004), "Talking Trash: Legitimacy, impression management, and unsystematic risk in the context of the natural environment", *Academy of Management Journal*, 47 (1): 93–103.

Bansal, P. and Roth, K. (2000), "Why Companies Go Green: A Model of Ecological Responsiveness", *Academy of Management*, 43 (4): 717–736.

Baruch, Y. and Holtom, B.C. (2008), "Survey response rate levels and trends in organizational research", *Human Relations*, 61 (8): 1139–1160.

Basu, K. and Palazzo, G. (2008), "Corporate Social Responsibility: A Process Model of Sensemaking", *Academy of Management Review*, 33 (1): 122–136.

References

Becker, K. (2008), "Unlearning as a driver of sustainable change and innovation: three Australian case studies", *International Journal of Technology Management*, 42 (1–2): 89–106.

Benyus, J.M. (1998), *Biomimicry: Innovation Inspired by Nature*, Quill William Morrow, New York.

Berkhout, F. (2011), "Eco-innovation: reflections on an evolving research agenda", *International Journal of Technology, Policy and Management*, 11 (3–4): 191–197.

Bernard, H.R. (2013), *Social research methods: Qualitative and quantitative approaches*, Sage Publications Inc., Thousand Oaks.

Berry, M.A. and Rondinelli, D.A. (1998), "Proactive corporate environmental management: A new industrial revolution", *Academy of Management Executive*, 12 (2): 38–50.

Berry, W.D. (1993), *Understanding Regression Assumptions, Quantitative Applications in the Social Sciences*, Vol. 92, Sage University Papers, Newbury Park.

Birkinshaw, J., Hamel, G. and Mol, Michael, J. (2008), "Management Innovation", *Academy of Management Review*, 33 (4): 825–845.

Bjørn, A. and Hauschild, M.Z. (2013), "Absolute versus Relative Environmental Sustainability. What can the Cradle-to-Cradle and Eco-efficiency Concepts Learn from Each Other?", *Journal of Industrial Ecology*, 17 (2): 321–332.

Blind, K. and Mangelsdorf, A. (2016), "Motives to standardize: Empirical evidence from Germany", *Technovation*, 48: 13–24.

Blomsma, F. and Brennan, G. (2017), "The Emergence of Circular Economy. A New Framing Around Prolonging Resource Productivity", *Journal of Industrial Ecology*, 21 (3): 603–614.

Bocken, N.M.P., Farracho, M., Bosworth, R. and Kemp, R. (2014), "The front-end of eco-innovation for eco-innovative small and medium sized companies", *Journal of Engineering and Technology Management*, 31: 43–57.

Bocken, N.M.P., Ritala, P. and Huotari, P. (2017), "The Circular Economy. Exploring the Introduction of the Concept Among S&P 500 Firms", *Journal of Industrial Ecology*, 21 (3): 487–490.

Boons, F. and Lüdeke-Freund, F. (2013), "Business models for sustainable innovation: State of the art and steps towards a research agenda", *Journal of Cleaner Production*, 45: 9–19.

Bozdogan, K., Deyst, J., Hoult, D. and Lucas, M. (1998), "Architectural innovation in product development through early supplier integration", *R&D Management*, 28 (3): 163–173.

Braungart, M. and McDonough, W. (Eds.) (2011), *Die nächste industrielle Revolution: Die Cradle to Cradle-Community*, 3rd ed., CEP Europäische Verlagsanstalt, Leipzig.

Braungart, M., McDonough, W. and Bollinger, A. (2007), "Cradle-to-cradle design: creating healthy emissions – a strategy for eco-effective product and system design", *Journal of Cleaner Production*, 15 (13–14): 1337–1348.

Breusch, T.S. and Pagan, A.R. (1979), "A simple test for heteroscedasticity and random coefficient variation", *Econometrica*, 47 (5): 1287–1294.

Bryman, A. and Bell, E. (2015), *Business Research Methods,* 4th ed., Oxford Univ. Press, Oxford.

Büschgens, T., Bausch, A. and Balkin, D.B. (2013), "Organizational Culture and Innovation: A Meta-Analytic Review", *Journal of Product Innovation Management*, 30 (4): 763–781.

Caiado, R.G.G., Freitas Dias, R. De, Mattos, L.V., Quelhas, O.L.G. and Leal Filho, W. (2017), "Towards sustainable development through the perspective of eco-efficiency – A systematic literature review", *Journal of Cleaner Production*, 165: 890–904.

Carrillo-Hermosilla, J., Del Río, P. and Könnölä, T. (2010), "Diversity of eco-innovations: Reflections from selected case studies", *Journal of Cleaner Production*, 18 (10–11): 1073–1083.

Carroll, A.B. (1998), "The Four Faces of Corporate Citizenship", *Business and Society Review*, 100 (1): 1–7.

Carroll, A.B. (1999), "Corporate Social Responsibility. Evolution of a Definitional Construct", *Business & Society*, 38 (3): 268–295.

Carroll, A.B. and Shabana, K.M. (2010), "The Business Case for Corporate Social Responsibility: A Review of Concepts, Research and Practice", *International Journal of Management Reviews*, 12 (1): 85–105.

Chandy, R.K. and Tellis, G.J. (1998), "Organizing for Radical Product Innovation: The Overlooked Role of Willingness to Cannibalize", *Journal of Marketing Research*, XXXV: 474–487.

Chappin, M.M.H., Cambré, B., Vermeulen, P.A.M. and Lozano, R. (2015), "Internalizing sustainable practices: a configurational approach on sustainable forest management of the Dutch wood trade and timber industry", *Journal of Cleaner Production*, 107: 760–774.

Chen, Y.-S. (2008), "The Driver of Green Innovation and Green Image – Green Core Competence", *Journal of Business Ethics*, 81 (3): 531–543.

Chesbrough, H.W. (2003), *Open Innovation: The New Imperative for Creating and Profiting from Technology*, Harvard Business School Press, Boston.

Child, J. (1973), "Predicting and Understanding Organization Structure", *Administrative Science Quarterly*, 1973: 168–185.

Christensen, C.M. (1992), "Exploring the Limits of the Technology S-Curve. Part I: Component Technologies", *Production and Operations Management*, 1 (4): 334–357.

Christmann, P. and Taylor, G. (2006), "Firm self-regulation through international certifiable standards: Determinants of symbolic versus substantive implementation", *Journal of International Business Studies*, 36 (6): 863–878.

Chun, R. and Davies, G. (2006), "The Influence of Corporate Character on Customers and Employees: Exploring Similarities and Differences", *Journal of the Academy of Marketing Science*, 34 (2): 138–146.

Cohen, J., Cohen, P., West, S.G. and Aiken, L.S. (2003), *Applied Multiple Regression/Correlation Analysis for the Behavioral Sciences*, 3rd ed., Lawrence Erlbaum Associates, Inc., Mahwah, New Jersey.

Cohen, W.M. and Levinthal, D.A. (1990), "Absorptive Capacity: A New Perspective on Learning and Innovation", *Administrative Science Quarterly*, 35 (1): 128–152.

Conway, J.M. and Huffcutt, A.I. (2003), "A Review and Evaluation of Exploratory Factor Analysis Practices in Organizational Research", *Organizational Research Methods*, 6 (2): 147–168.

Cooper, R.G. (1988), "Predevelopment activities determine new product success", *Industrial Marketing Management*, 17 (3): 237–247.

Cooper, R.G. and Kleinschmidt, E.J. (1987), "New products: What separates winners from losers?", *Journal of Product Innovation Management*, 4 (3): 169–184.

Cooper, R.G. and Kleinschmidt, E.J. (1991), "New product processes at leading industrial firms", *Industrial Marketing Management*, 20 (2): 137–147.

Cradle to Cradle Products Innovation Institute (2013), *Innovations Stories*, C2C PII.

Cradle to Cradle Products Innovation Institute (2014), *Pilot study: Impacts of the Cradle to Cradle certified products program. Technical Report*, C2C PII, available at: https://www.c2ccertified.org/impact-study#:~:text=Rethinking%20production%20and%20design%20processes,water%20and%20energy%20efficiency%20improvements (accessed 19 April 2019)

Cradle to Cradle Products Innovation Institute (2019a), "C2C Certified Products Registry", available at: https://www.c2ccertified.org/products/registry (accessed 19 April 2019).

Cradle to Cradle Products Innovation Institute (2019b), "Find an Assessor", available at: https://www.c2ccertified.org/get-certified/find-an-assessor (accessed 19 April 2019).

Cradle to Cradle Products Innovation Institute (2019c), "Resources", available at: https://www.c2ccertified.org/resources/detail/cradle-to-cradle-certified-fees-schedule (accessed 3 April 2019).

Cradle to Cradle Products Innovation Institute (2020), "How to certify", available at: https://www.c2ccertified.org/get-certified/product-certification-process (accessed 20 December 2020).

Creswell, J.W. (2009), *Research Design: Qualitative, Quantitative, and Mixed Methods Approaches,* 3rd ed., Sage Publications Inc, Thousand Oaks.

Crilly, D., Zollo, M. and Hansen, M.T. (2012), "Faking It or Muddling Through? Understanding Decoupling in Response to Stakeholder Pressures", *Academy of Management Journal,* 55 (6): 1429–1448.

Cronbach, L.J. (1951), "Coefficient alpha and the internal structure of tests", *Psychometrika,* 16 (3): 297–334.

Daly, H.E. (1997), *Beyond Growth: The economics of sustainable development,* Beacon Press, Boston.

Damanpour, F. (1992), "Organizational Size and Innovation", *Organization Studies,* 13 (3): 375–402.

Danneels, E. (2008), "Organizational antecedents of second-order competences", *Strategic Management Journal,* 29 (5): 519–543.

Danneels, E. and Kleinschmidt, E.J. (2001), "Product innovativeness from the firm's perspective: Its dimensions and their relation with project selection and performance", *Journal of Product Innovation Management,* 18: 357–373.

Dawson, J.F. (2014), "Moderation in Management Research: What, Why, When, and How", *Journal of Business and Psychology,* 29 (1): 1–19.

De Pauw, I.C., Karana, E. and Kandachar, P. (2013), "Cradle to Cradle in Product Development: A Case Study of Closed-Loop Design", in Nee A., Song B. and Ong S. K. (Eds.), *Re-engineering Manufacturing for Sustainability,* Springer, Singapore: 47–52.

Del Río, P., Carrillo-Hermosilla, J. and Könnölä, T. (2010), "Policy Strategies to Promote Eco-Innovation", *Journal of Industrial Ecology,* 14 (4): 541–557.

Delmas, M.A. and Burbano, V.C. (2011), "The Drivers of Greenwashing", *California Management Review,* 54 (1): 64–87.

Delmas, M.A., Nairn-Birch, N. and Balzarova, M. (2013), "Choosing the Right Eco-Label for Your Product", *MIT Sloan Management Review,* 54 (4): 10–12.

Den Hollander, M.C., Bakker, C.A. and Hultink, E.J. (2017), "Product Design in a Circular Economy. Development of a Typology of Key Concepts and Terms", *Journal of Industrial Ecology,* 21 (3): 517–523.

Diaz Lopez, F.J., Bastein, T. and Tukker, A. (2018), "Business Model Innovation for Resource-efficiency, Circularity and Cleaner Production. What 143 Cases Tell Us", *Ecological Economics,* 155: 20–35.

Doty, D.H. and Glick, W.H. (1998), "Common Methods Bias: Does Common Methods Variance Really Bias Results?", *Organizational Research Methods,* 1 (4): 374–406.

Dziuban, C.D. and Shirkey, E.C. (1974), "When is a correlation matrix appropriate for factor analysis? Some decision rules", *Psychological Bulletin,* 81 (6): 358–361.

References

Edmondson, A.C. and McManus, S.E. (2007), "Methodological Fit in Management Field Research", *Academy of Management Review*, 32 (4): 1155–1179.

Ehrenfeld, J.R. (1997), "Industrial ecology: a framework for product and process design", *Journal of Cleaner Production*, 5 (1–2): 87–95.

Elkington, J. (1994), "Towards the Sustainable Corporation: Win-Win-Win Business Strategies for Sustainable Development", *California Management Review*: 90–100.

Elkington, J. (1997), *Cannibals with Forks: The Triple Bottom Line of 21st Century Business*, Capstone Publishing Ltd, Oxford.

Ellen MacArthur Foundation (2012), *Towards a Circular Economy: Economic and business rationale for an accelerated transition*, Ellen MacArthur Foundation.

Ellen MacArthur Foundation (2013), *Towards the Circular Economy: Opportunities for the consumer goods sector*, Ellen MacArthur Foundation.

Ellen MacArthur Foundation (2015a), *Growth Within: A Circular Economy Vision for a Competitive Europe*, Ellen MacArthur Foundation.

Ellen MacArthur Foundation (2015b), *Towards a Circular Economy: Business Rationale for an accelerated transition*, Ellen MacArthur Foundation.

Ellen MacArthur Foundation (2016a), "Our Story", available at: https://www.ellenmacarthurfoundation.org/our-story/mission (accessed 13 January 2016).

Ellen MacArthur Foundation (2016b), *Intelligent Assets: Unlocking the Circular Economy Potential*.

Esty, D.C. and Porter, M.E. (1998), "Industrial Ecology and Competitiveness. Strategic Implications for the Firm", *Journal of Industrial Ecology*, 2 (1): 35–43.

European Commission (2020), "Circular economy", available at: https://ec.europa.eu/environment/circular-economy/first_circular_economy_action_plan.html (accessed 22 November 2020).

Evanschitzky, H., Eisend, M., Calantone, R.J. and Jiang, Y. (2012), "Success Factors of Product Innovation. An Updated Meta-Analysis", *Journal of Product Innovation Management*, 29 (3): 21–37.

Fink, A. (2013), *How to conduct surveys: A step-by-step guide*, 5th ed., Sage Publications Inc., Thousand Oaks.

Foster, R.N. (1986), *Innovation: The Attacker's Advantage*, Simon and Schuster, New York.

Fuller, C.M., Simmering, M.J., Atinc, G., Atinc, Y. and Babin, B.J. (2016), "Common methods variance detection in business research", *Journal of Business Research*, 69 (8): 3192–3198.

Geissdoerfer, M., Savaget, P., Bocken, N.M.P. and Hultink, E.J. (2016), "The Circular Economy – A new sustainability paradigm?", *Journal of Cleaner Production*, 143: 757–768.

Geng, V. and Herstatt, C. (2014), "The cradle-to-cradle (C2C) paradigm in the context of innovation management and driving forces for implementation", *Technology and Innovations Management Working Paper No. 79*.

Gerbing, D.W. and Anderson, J.C. (1988), "An Updated Paradigm for Scale Development Incorporating Unidimentionality and its Assessment", *Journal of Marketing Research*, 25: 186–192.

Ghisellini, P., Cialani, C. and Ulgiati, S. (2016), "A review on circular economy: The expected transition to a balanced interplay of environmental and economic systems", *Journal of Cleaner Production*, 114: 11–32.

Gioia, D.A., Corley, K.G. and Hamilton, A.L. (2013), "Seeking Qualitative Rigor in Inductive Research: Notes on the Gioia Methodology", *Organizational Research Methods*, 16 (1): 15–31.

Golden, B.R. (1992), "The past is the past - Or is it? The use of retrospective accounts as indicators of past strategy", *Academy of Management Journal*, 35 (4): 848–860.

Gray, R., Kouhy, R. and Lavers, S. (1995), "Corporate social and environmental reporting. A review of the literature and a longitudinal study of UK disclosure", *Accounting, Auditing & Accountability Journal*, 8 (2): 47–77.

Greening, L.A., Greene, D.L. and Difiglio, C. (2000), "Energy efficiency and consumption - the rebound effect – a survey", *Energy Policy*, 28: 389–401.

Guide, V.D.R.J. and Van Wassenhove, L.N. (2009), "The Evolution of Closed-Loop Supply Chain Research", *Operations Research*, 57 (1): 10–18.

Hahn, T., Preuss, L., Pinkse, J. and Figge, F. (2015), "Cognitive frames in corporate sustainability: Managerial sensemaking with paradoxical and business case frames", *Academy of Management Review*, 39 (4): 18–42.

Hair, J.F., Black, W.C., Babin, B.J. and Anderson, R.E. (2014), *Multivariate Data Analysis*, 7th ed., Pearson Education Limited, Harlow.

Hall, J. and Vredenburg, H. (2003), "The Challenges of Innovating for Sustainable Development.", *MIT Sloan Management Review*, Fall: 61–68.

Hannan, T.H. and McDowell, J.M. (1984), "The determinants of technology adoption: The case of the banking firm", *The RAND Journal of Economics*, 1984: 328–335.

Hansen, E. G., Grosse-Dunker, F. and Reichwald, R. (2009), "Sustainability Innovation Cube. A framework to evaluate sustainability-oriented innovations", *International Journal of Innovation Management*, 13 (4): 683–713.

Hansen, E. G. and Schaltegger, S. (2016), "The Sustainability Balanced Scorecard. A Systematic Review of Architectures", *Journal of Business Ethics*, 133 (2): 193–221.

Hansen, E. G. and Schmitt, J. (2020), "Orchestrating Cradle-to-Cradle Innovation Across the Value Chain: Overcoming Barriers through Innovation Communities, Collaboration Mechanisms, and Intermediation", *Journal of Industrial Ecology*, 21 (4): 938–952.

Hansen, E. G., Wicki, S. and Schaltegger, S. (2018), "Structural ambidexterity, transition processes, and integration trade-offs: a longitudinal study of failed exploration", *R&D Management*, 49 (4): 484–508.

Hart, S. (1993), "Dimensions of Success in New Product Development: an Exploratory Investigation", *Journal of Marketing Management*, 9 (1): 23–41.

Hayes, A.F. (2013), *Introduction to mediation, moderation, and conditional process analysis: A regression-based approach*, The Guilford Press, New York.

Helms, W.S., Oliver, C. and Webb, K. (2012), "Antecedents of settlement on a new institutional practice: Negotiation of the ISO 26000 standard on social responsibility", *Academy of Management Journal*, 55 (5): 1120–1145.

Henderson, R.M. and Clark, K.B. (1990), "Architectural Innovation. The reconfiguration of existing product technologies and the failure of established firms", *Administrative Science Quarterly*, 35: 9–30.

Henson, R.K. (2006), "Use of Exploratory Factor Analysis in Published Research: Common Errors and Some Comment on Improved Practice", *Educational and Psychological Measurement*, 66 (3): 393–416.

Hirsch, P.M. and Levin, D.Z. (1999), "Umbrella Advocates Versus Validity Police: A Life-Cycle Model", *Organization Science*, 10 (2): 199–212.

Horbach, J., Rammer, C. and Rennings, K. (2012), "Determinants of eco-innovations by type of environmental impact. The role of regulatory push/pull, technology push and market pull", *Ecological Economics*, 78: 112–122.

Huber, J. (1995), *Nachhaltige Entwicklung: Strategien für eine ökologische und soziale Erdpolitik*, Ed. Sigma, Berlin.

Hurley, A.E., Scandura, T.A., Schriesheim, C.A., Brannick, M.T., Seers, A., Vandenberg, R.J. and Williams, L.J. (1997), "Exploratory and confirmatory factor analysis: guidelines, issues, and alternatives", *Journal of Organizational Behavior*, 18: 667–683.

Iarossi, G. (2006), *The Power of Survey Design: A User's Guide for Managing Surveys, Interpreting Results, and Influencing Respondents*, World Bank, Washington, D.C.

Jayaraman, V. and Luo, Y. (2007), "Creating Competitive Advantages through new value creation: A reverse logistics perspective", *Academy of Management Perspectives*, 21 (2): 56–73.

Jennrich, R.I. (2006), "Rotation to Simple Loadings Using Component Loss Functions: The Oblique Case", *Psychometrika*, 71 (1): 173–191.

Johansson, G. (2002), "Success factors for integration of ecodesign in product development. A review of state of the art", *Environmental Management and Health*, 13 (1): 98–107.

Kaiser, H.F. and Rice, J. (1974), "Little Jiffy, Mark IV", *Educational and Psychological Measurement*, 34: 111–117.

Kalogerakis, K., Drabe, V., Paramasivam, M. and Herstatt, C. (2015), "Closed-Loop Supply Chains for Cradle to Cradle Products", in Kersten, W., Blecker, T. and Ringle, C.M. (Eds.), *Sustainability in Logistics and Supply Chain Management: New Designs and Strategies*, Techn. Univ. Hamburg-Harburg Univ. Bibl; Epubli, Hamburg, Berlin: 3–34.

Kausch, M.F. and Klosterhaus, S. (2015), "Response to 'Are Cradle to Cradle certified products environmentally preferable? Analysis from an LCA approach'", *Journal of Cleaner Production*, 113: 715–716.

Kennedy, M.T. and Fiss, P.C. (2009), "Institutionalization, framing, and diffusion: The logic of TQM adoption and implementation decisions among U.S. hospitals", *Academy of Management Journal*, 52 (5): 897–918.

Ketata, I., Sofka, W. and Grimpe, C. (2015), "The role of internal capabilities and firms environment for sustainable innovation: evidence for Germany", *R&D Management*, 45 (1): 60–75.

Kimberly, J.R. (1976), "Organizational Size and the Structuralist Perspective: A Review, Critique, and Proposal", *Administrative Science Quarterly*, 21 (December): 571–597.

Kirchherr, J., Piscicelli, L., Bour, R., Kostense-Smit, E., Muller, J., Huibrechtse-Truijens, A. and Hekkert, M. (2018), "Barriers to the Circular Economy: Evidence From the European Union (EU)", *Ecological Economics*, 150 (2018): 264–272.

Kirchherr, J., Reike, D. and Hekkert, M. (2017), "Conceptualizing the circular economy. An analysis of 114 definitions", *Resources, Conservation and Recycling*, 127: 221–232.

Kobayashi, Y., Kobayashi, H., Hongu, A. and Sanehira, K. (2006), "A Practical Method for Quantifying Eco-efficiency Using Eco-design Support Tools", *Journal of Industrial Ecology*, 9 (4): 131–144.

Koen, P.A., Ajamian, G., Boyce, S., Clamen, A., Fisher, E., Fountoulakis, S., Johnson, A., Pushpinder, P. and Seibert, R. (2002), "Fuzzy Front End: Effective Methods, Tools and Techniques", in Belliveau, P., Griffin, A. and Somermeyer S. (Eds), *The PDMA Toolbook for New Product Development*, John Wiley & Sons, New York.

Kotter, J.P. (1995), "Leading Change: Why transformation efforts fail", *Harvard Business Review*, 1995 (March-April): 59–87.

Kuckartz, U., Ebert, T., Rädiker, S. and Stefer, C. (2009), *Evaluation online: Internetgestützte Befragung in der Praxis*, VS Verlag für Sozialwissenschaften, Wiesbaden.

Kumar, N., Stern, L.W. and Anderson, J.C. (1993), "Conducting interorganizational research using key informants", *Academy of Management Journal*, 36 (6): 1633–1651.

Ladhari, R. (2010), "Developing e-service quality scales. A literature review", *Journal of Retailing and Consumer Services*, 17 (6): 464–477.

LeGwin, J. (2000), "A Theoretical Foundation for Life-Cycle Assessment", *Journal of Industrial Ecology*, 4 (1): 13–28.

References

Lieder, M. and Rashid, A. (2016), "Towards Circular Economy implementation: A comprehensive review in context of manufacturing industry", *Journal of Cleaner Production*, 115: 36-51.

Little, R.J.A. and Rubin, D.B. (2002), *Statistical analysis with missing data*, Wiley Series in Probability and Statistics, 2nd ed., Wiley, Hoboken.

Llorach-Massana, P., Farreny, R. and Oliver-Solà, J. (2015), "Are Cradle to Cradle certified products environmentally preferable? Analysis from an LCA approach", *Journal of Cleaner Production*, 93: 243–250.

Lynn, G.S., Abel, K.D., Valentine, W.S. and Wright, R.C. (1999), "Key Factors in Increasing Speed to Market and Improving New Product Success Rates", *Industrial Marketing Management*, 28 (4): 319–326.

MacCallum, R.C. and Widaman, K.F. (1999), "Sample Size in Factor Analysis", *Psychological Methods*, 4 (1): 84–99.

Margolis, J.D. and Walsh, J.P. (2003), "Misery Loves Companies: Rethinking Social Initiatives by Business", *Administrative Science Quarterly*, 48: 268–305.

Martín-Peña, M.L., Díaz-Garrido, E. and Sánchez-López, J.M. (2014), "Analysis of benefits and difficulties associated with firms' Environmental Management Systems. The case of the Spanish automotive industry", *Journal of Cleaner Production*, 70: 220–230.

McDonough Braungart Design Chemistry, LLC (2012), "Overview of the Cradle to Cradle Certified Product Standard", available at: https://www.c2ccertified.org/images/uploads/ C2CCertified_V3_Overview_121113.pdf (accessed 4 June 2014).

McDonough, W. and Braungart, M. (2002a), "Design for the Triple Top Line: New Tools for Sustainable Commerce", *Corporate Environmental Strategy*, 9 (3): 251–258.

McDonough, W. and Braungart, M. (2002b), *Remaking the way we make things: Cradle to cradle*, North Point Press, New York.

McDonough, W. and Braungart, M. (2012), *Celebrating 20 years: The Hannover Principles: Design for Sustainability*, available at: https://mcdonough.com/wp-content/uploads/2013/03/Hannover-Principles-1992.pdf (accessed 22 January 2015)

McDonough, W. and Braungart, M. (2013), *The Upcycle: Beyond Sustainability – Designing for Abundance*, Melcher Media, New York.

McDonough, W., Braungart, M., Anastas, P.T. and Zimmerman, J.B. (2003), "Applying the Principles of Green Engineering to Cradle-to-Cradle Design", *Environmental Science & Technology*, 2003: 434–441.

Mendoza, J.M.F., Sharmina, M., Gallego-Schmid, A., Heyes, G. and Azapagic, A. (2017), "Integrating Backcasting and Eco-Design for the Circular Economy. The BECE Framework", *Journal of Industrial Ecology*, 21 (3): 526–544.

Miles, M.B. and Huberman, A.M. (1994), *Qualitative Data Analysis*, 2nd ed., Sage Publications Inc., Thousand Oaks.

Miles, M.B., Huberman, A.M. and Saldana, J. (2014), *Qualitative Data Analysis: A Methods Sourcebook,* 3rd ed., Sage Publications Inc., Thousand Oaks.

Morana, R. and Seuring, S. (2007), "End-of-life returns of long-lived products from end customer—insights from an ideally set up closed-loop supply chain", *International Journal of Production Research*, 45 (18–19): 4423–4437.

Neutzling, D.M., Land, A., Seuring, S. and Nascimento, L.F.M.d. (2018), "Linking sustainability-oriented innovation to supply chain relationship integration", *Journal of Cleaner Production*, 172: 3448–3458.

Nidumolu, R., Prahalad C.K. and Rangaswami, M.R. (2009), "Why Sustainability Is Now the Key Driver of Innovation", *Harvard Business Review*, 87 (9): 56–64.

Niero, M., Hauschild, M.Z., Hoffmeyer, S.B. and Olsen, S.I. (2017), "Combining Eco-Efficiency and Eco-Effectiveness for Continuous Loop Beverage Packaging Systems. Lessons from the Carlsberg Circular Community", *Journal of Industrial Ecology*, 21 (3): 742–753.

Nunnally, J. and Bernstein, I. (1994), *Psychometric Theory,* 3rd ed., McGraw-Hill, New York.

Östlin, J., Sundin, E. and Björkman, M. (2008), "Importance of closed-loop supply chain relationships for product remanufacturing", *International Journal of Production Economics*, 115 (2): 336–348.

Paramanathan, S., Farrukh, C., Phaal, R. and Probert, D. (2004), "Implementing industrial sustainability: the research issues in technology management", *R&D Management*, 34 (5): 527–537.

Pavlou, P.A., Liang, H. and Xue, Y. (2007), "Understanding and mitigating uncertainty in online exchange relationships: A principal-agent perspective", *MIS Quarterly*, 31 (1): 105–136.

Petersen, M. and Brockhaus, S. (2017), "Dancing in the dark. Challenges for product developers to improve and communicate product sustainability", *Journal of Cleaner Production*, 161: 345–354.

Peterson, R.A. (2000), *Constructing effective questionnaires*, Sage Publications Inc., Thousand Oaks.

Podsakoff, P.M., MacKenzie, S.B., Lee, J.-Y. and Podsakoff, N.P. (2003), "Common method biases in behavioral research. A critical review of the literature and recommended remedies", *The Journal of Applied Psychology*, 88 (5): 879–903.

Podsakoff, P.M., MacKenzie, S.B. and Podsakoff, N.P. (2012), "Sources of method bias in social science research and recommendations on how to control it", *Annual Review of Psychology*, 63: 539–569.

Porter, M.E. (1985), *Competitive Advantage: Creating and Sustaining Superior Performance*, The Free Press, New York.

Porter, M.E. and Kramer, M.R. (2006), "Strategy and Society: The Link Between Competitive Advantage and Corporate Social Responsibility", *Harvard Business Review*, 84 (12): 78–94.

Porter, M.E. and Van der Linde, C. (1995), "Toward a New Conception of the Environment-Competitiveness Relationship", *Journal of Economic Perspectives*, 9 (4): 97–118.

Potthoff, R.F. (1964), "On the Johnson-Neyman technique and some extensions thereof", *Psychometrika*, 29 (3): 241–256.

Punch, K.F. (2014), *Introduction to social research: quantitative and qualitative approaches*, 3rd ed., Los Angeles.

Reay, S.D., McCool, J.R. and Withell, A. (2011), "Exploring the Feasibility of Cradle to Cradle (Product) Design: Perspectives from New Zealand Scientists", *Journal of Sustainable Development*, 4 (1): 36–44.

Reijnders, L. (1998), "The Factor X Debate: Setting Targets for Eco-Efficiency", *Journal of Industrial Ecology*, 2 (1): 13–22.

Reijnders, L. (2008), "Are emissions or wastes consisting of biological nutrients good or healthy?", *Journal of Cleaner Production*, 16 (10): 1138–1141.

Rindfleisch, A. and Moorman, C. (2001), "The Acquisition and Utilization of Information in New Product Alliances: A Strength-of-Ties Perspective", *Journal of Marketing*, 65 (2): 1–18.

Robinson, J.P., Shaver, P.R. and Wrightsman, L.S. (1991), "Criteria for Scale Selection and Evaluation", in Robinson, J.P. (Ed.), *Measures of personality and social psychological attitudes, Measures of social psychological attitudes*, Academic Press, San Diego: 1–16.

Rossi, M., Charon, M., Wing, G. and Ewell, J. (2006), "Design for the Next Generation: Incorporating Cradle to Cradle Design into Herman Miller Products", *Journal of Industrial Ecology*, 10 (4): 193–210.

Rubik, F., Scheer, D. and Iraldo, F. (2008), "Eco-labelling and product development: potentials and experiences", *International Journal of Product Development*, 6 (3–4): 393–419.

Saldana, J. (2013), *The Coding Manual for Qualitative Researchers*, 2nd ed., Sage Publications Inc, London.

Schaltegger, S., Beckmann, M. and Hansen, E. G. (2013), "Transdisciplinarity in Corporate Sustainability. Mapping the Field", *Business Strategy and the Environment*, 22 (4): 219–229.

Schaltegger, S. and Hörisch, J. (2017), "In Search of the Dominant Rationale in Sustainability Management. Legitimacy- or Profit-Seeking?", *Journal of Business Ethics*, 145 (2): 259–276.

Schaltegger, S. and Synnestvedt, T. (2002), "The link between 'green' and economic success: environmental management as the crucial trigger between environmental and economic performance", *Journal of Environmental Management*, 65 (4): 339–346.

Schein, E.H. (2010), *Organizational Culture and Leadership*, 4th ed., Jossey-Bass, San Francisco.

Schiederig, T., Tietze, F. and Herstatt, C. (2012), "Green innovation in technology and innovation management an exploratory literature review", *R&D Management*, 42 (2): 180–192.

Schmidheiny, S. (1992), *Changing course: A global business perspective on development and the environment*, MIT press, Cambridge, Massachusetts.

Schmitt, J. and Hansen, E. G. (2017), "Promoting circular innovation through innovation networks: the case of cradle to cradle certified products", paper presented at PLATE conference, 8-10 November 2017, Amsterdam.

Schmitt, J. and Hansen, E. G. (2018), "Circular Innovation Processes from an Absorptive Capacity Perspective. The Case of Cradle to Cradle", *Academy of Management Proceedings*, 2018: 1–40.

Schons, L. and Steinmeier, M. (2016), "Walk the Talk? How Symbolic and Substantive CSR Actions Affect Firm Performance Depending on Stakeholder Proximity", *Corporate Social Responsibility and Environmental Management*, 23 (6): 358–372.

Seebode, D., Jeanrenaud, S. and Bessant, J. (2012), "Managing innovation for sustainability", *R&D Management*, 42 (3): 195–206.

Seidler, J. (1974), "On Using Informants: A Technique for Collecting Quantitative Data and Controlling for Measurement Error in Organization Analysis", *American Sociological Review*, 39: 816–831.

Sekaran, U. and Bougie, R. (2010), *Research Methods for Business: A Skill Building Approach,* 5th ed., John Wiley & Sons.

Senge, P. (Ed.) (2008), *The necessary revolution: Working together to create a sustainable world*, Broadway Books, New York.

Simpson, D., Power, D. and Klassen, R. (2012), "When One Size Does Not Fit All. A Problem of Fit Rather than Failure for Voluntary Management Standards", *Journal of Business Ethics*, 110 (1): 85–95.

Smits, A., Drabe, V. and Herstatt, C. (2020), "Beyond motives to adopt. Implementation configurations and implementation extensiveness of a voluntary sustainability standard", *Journal of Cleaner Production*, 251: 1–15.

Song, M. and Montoya-Weiss, M.M. (2001), "The effect of perceived technological uncertainty on Japanese New Product Development", *Academy of Management Journal*, 44 (1): 61–80.

Spector, P.E. (2006), "Method Variance in Organizational Research. Truth or Urban Legend?", *Organizational Research Methods*, 9 (2): 221–232.

Stahel, W.R. (1982), *The Product-Life Factor, Series: 1982 Mitchell Prize Papers*, NARC: 72–96.

Stahel, W.R. (2010), *The Performance Economy*, Palgrave Macmillan, Basingstoke.

Statista (2020a), "Global market value of plastic", available at: https://www.statista.com/statistics/1060583/global-market-value-of-plastic/(accessed 22 November 2020).

Statista (2020b), "Plastic waste in Europe", available at: https://www.statista.com/topics/5141/plastic-waste-in-europe/= (accessed 22 November 2020).

Stevens, J. (2002), *Applied multivariate statistics for the social sciences*, 4th ed., Lawrence Erlbaum Associates, Inc., Mahwah, New Jersey.

Stock, T., Obenaus, M., Slaymaker, A. and Seliger, G. (2017), "A Model for the Development of Sustainable Innovations for the Early Phase of the Innovation Process", *Procedia Manufacturing*, 8: 215–222.

Tollin, K. and Vej, J. (2012), "Sustainability in business: understanding meanings, triggers and enablers", *Journal of Strategic Marketing*, 20 (7): 625–641.

Tourangeau, R., Rips, L.J. and Rasinski, K. (2000), *The Psychology of Survey Response*, Cambridge University Press, Cambridge.

Toxopeus, M.E., De Koeijer, B.L.A. and Meij, A.G.G.H. (2015), "Cradle to Cradle. Effective Vision vs. Efficient Practice?", *Procedia CIRP*, 29: 384–389.

Tracy, S.J. (2013), *Qualitative Research Methods: Collecting evidence, crafting analysis, communicating impact*, Wiley-Blackwell, Hoboken, New Jersey.

Tukker, A. (2004), "Eight types of product–service system: eight ways to sustainability? Experiences from SusProNet", *Business Strategy and the Environment*, 13 (4): 246–260.

Tukker, A. (2015), "Product services for a resource-efficient and circular economy – a review", *Journal of Cleaner Production*, 97: 76–91.

United Nations (1987), *Report of the World Commission on Environment and Development: Our Common Future*, World Commission on Environment and Development.

Vaccaro, I.G., Jansen, J.J.P., Van den Bosch, F.A.J. and Volberda, H.W. (2012), "Management Innovation and Leadership. The Moderating Role of Organizational Size", *Journal of Management Studies*, 49 (1): 28–51.

Van de Ven, A.H. and Polley, D. (1992), "Learning while innovating", *Organization Science*, 3 (1): 92–116.

Van Marrewijk, M. and Werre, M. (2003), "Multiple Levels of Corporate Sustainability", *Journal of Business Ethics*, 44: 107–119.

Vanpoucke, E., Vereecke, A. and Wetzels, M. (2014), "Developing supplier integration capabilities for sustainable competitive advantage. A dynamic capabilities approach", *Journal of Operations Management*, 32 (7–8): 446–461.

Verworn, B. (2004), *Die frühen Phasen der Produktentwicklung: Eine empirische Analyse in der Mess-, Steuer- und Regelungstechnik*, Deutscher Universitätsverlag, Wiesbaden.

Weiber, R. and Mühlhaus, D. (2014), *Strukturgleichungsmodellierung: Eine anwendungsorientierte Einführung in die Kausalanalyse mit Hilfe von AMOS, SmartPLS und SPSS, Springer-Lehrbuch,* 2nd ed., Springer Gabler, Berlin.

Wheelwright, S.C. and Clark, K.B. (1992), "Creating project plans to focus product development", *Harvard Business Review*, 1992: 70–82.

Wheelwright, S.C. and Clark, K.B. (1995), *Leading Product Development: The Senior Manager's Guide to Creating and Shaping the Enterprise*, The Free Press, New York.

Williams, B., Brown, T. and Onsman, A. (2010), "Exploratory factor analysis: A five-step guide for novices", *Journal of Emergency Primary Health Care*, 8 (3): 1–13.

Wolff, H.-G. and Bacher, J. (Eds.) (2010), *Hauptkomponentenanalyse und explorative Faktorenanalyse*, VS Verlag für Sozialwissenschaften, Wiesbaden.

Wolfinbarger, M. and Gilly, M.C. (2003), "eTailQ: dimensionalizing, measuring and predicting etail quality", *Journal of Retailing*, 79 (3): 183–198.

World Economic Forum (2020), "Circular Economy and Material Value Chains", available at: https://www.weforum.org/projects/circular-economy (accessed 22 November 2020).

Zahra, S.A. and George, G. (2002), "Absorptive Capacity: A review, reconceptualization, and extension", *Academy of Management Review*, 27 (2): 185–203.

Zink, T. and Geyer, R. (2017), "Circular Economy Rebound", *Journal of Industrial Ecology*, 21 (3): 593–602.

8 Appendix

8.1 Appendix A: Online survey

Technology and Innovation Management at Hamburg University of Technology

Cradle to Cradle - Motivators and enabling conditions for implementation

Dear participant,

this questionnaire is designed to study motivational factors for the implementation of the Cradle to Cradle (C2C) concept and to look at the conditions that facilitate or hinder the implementation of C2C in companies.

We kindly ask you to participate in the survey as your company belongs to the distinctive group of companies which have valuable experience in implementing C2C, no matter whether you currently hold a certificate or not. The information you provide will help us to better understand the implementation process of C2C and derive helpful implications that can be useful for further research and for managerial practice. Please give the answers to the best of your knowledge and judgement from the perspective of your company.

This research activity is conducted in the context of a research project at the Institute for Technology and Innovation Management at Hamburg University of Technology and is independent of the work of EPEA or MBDC (McDonough Braungart Design Chemistry). Your personal and individual company data and the single survey answers will be handled with strict confidentiality.

We would be happy to share with you the analysis results (no individual company name will be published). If you are interested in the results, please indicate your e-mail address after completing the survey.

We greatly appreciate your time and cooperation. Thank you.

Appendix

Cradle to Cradle - Motivators and enabling conditions for implementation

To your knowledge, does your company currently hold a C2C certificate for one or more products from your product portfolio?

○ Yes
○ No
○ I don't know

When has your company received the first C2C certification?

○ Before 2008   ○ 2010   ○ 2013
○ 2008   ○ 2011   ○ 2014
○ 2009   ○ 2012   ○ 2015

To your knowledge, what is the share (approximately) in the total number of products of your company's C2C certified products? Please do not consider the products that are currently in the certification process.

○ Below 10%
○ 10-30%
○ 31-50%
○ 51-70%
○ More than 70%
○ N/A

To your knowledge, what is the share (approximately) in overall turnover for your company's C2C certified products?

○ Below 10%
○ 10-30%
○ 31-50%
○ 51-70%
○ More than 70%
○ N/A

Appendix A: Online survey

To your knowledge, how many products from your company's product portfolio are currently in the process of getting C2C certification? (An approximate number is sufficient)

Approximate number of products: [   ]

Please indicate the highest C2C certification level that one (or more) of your products achieved.

- ○ Basic
- ○ Bronze
- ○ Silver
- ○ Gold
- ○ Platinum
- ○ N/A

Please indicate the C2C certification level that most of your company's certified products have.

- ○ Basic
- ○ Bronze
- ○ Silver
- ○ Gold
- ○ Platinum
- ○ N/A

Did your company increase the number of C2C certified products after receiving the first certificate(s)?

- ○ Yes
- ○ No
- ○ N/A

Is your company planning to *prolong* the C2C certification in the future?

- ○ Yes, definitely
- ○ Maybe
- ○ No, I don't think so

Is your company planning to *extend* the C2C certification to other products in the future?

- ○ Yes, definitely
- ○ Maybe
- ○ No, I don't think so

Appendix

When did the last C2C certificate expire?

- ○ Before 2010
- ○ 2010
- ○ 2011
- ○ 2012
- ○ 2013
- ○ 2014
- ○ 2015
- ○ N/A

What were the main reasons to not prolong the C2C certification?

- ○ The customer reaction was not as positive as expected.
- ○ The cost for the certification was too high compared to the benefit.
- ○ The cost of implementation was too high compared to the benefit.
- ○ The suppliers were not able to provide the demanded material.
- ○ Other (please specify)

Does your company hold other certificates/standards/labels with regards to environmental or social responsibility?

- ○ Yes
- ○ No
- ○ N/A

Please check all applicable labels and feel free to add the ones that are not listed but do apply to your company

- ☐ Energy Star
- ☐ ISO standard(s)
- ☐ LEED – US Green Building Council
- ☐ FSC – Forest Stewardship Council
- ☐ Der Blaue Engel (Blue Angel)
- ☐ Öko-Tex
- ☐ EMAS
- ☐ Fairtrade
- ☐ Other (please specify)

Appendix A: Online survey

Please select the level of importance of the following reasons for your company's decision to start implementing C2C standards and certification:

	Unimportant	Of little importance	Moderately important	Important	Very important	N/A
The fit of the C2C concept with the company's philosophy	○	○	○	○	○	○
The opportunity to get the company's sustainability efforts approved by an independent certification institute	○	○	○	○	○	○
The company CEO's / top management's strong endeavor to achieve the C2C certification	○	○	○	○	○	○
The expectation to contribute to the company's cost reduction	○	○	○	○	○	○
The expectation to contribute to the company's risk reduction	○	○	○	○	○	○
The expectation to increase sales	○	○	○	○	○	○
The expectation of C2C to be a source of new opportunities	○	○	○	○	○	○
The potential loss of market share if the company does not implement C2C standards	○	○	○	○	○	○
The competition from other C2C certified companies	○	○	○	○	○	○
The expectation to improve the quality of the company's product(s)	○	○	○	○	○	○
The expectation to be perceived as a market leader	○	○	○	○	○	○
The expectation to improve customer satisfaction	○	○	○	○	○	○
The demand of the company's customers for C2C	○	○	○	○	○	○
The expectation to better comply with current legislation	○	○	○	○	○	○

Other (please specify)

Please indicate your agreement on the following statements

	Strongly disagree	Disagree	Neither nor	Agree	Strongly agree	N/A
The company has integrated the C2C standards in procedures and work instructions	○	○	○	○	○	○
The company has identified specific persons and positions responsible for C2C implementation	○	○	○	○	○	○
The company has adapted the C2C implementation procedures to its various business departments, business units or plants/warehouses	○	○	○	○	○	○
The company has integrated the C2C standards in its computerized and other administrative systems	○	○	○	○	○	○
The company keeps records of the training provided to staff in relation to the implementation of C2C standards	○	○	○	○	○	○
The company has obliged its supply base to supply according to C2C standards	○	○	○	○	○	○

Other (please specify)

Appendix

Please indicate the extent to which you believe that at this point in time C2C philosophy, standards, and methods have been implemented throughout your company

Not at all implemented		Somehow implemented		To a great extent implemented	N/A
○	○	○	○	○	○

Other (please specify)

Please rate the degree to which the following statements describe the nature of your company's overall relationship with EPEA and/or MBDC, suppliers and other partners

	Strongly disagree	Disagree	Neither nor	Agree	Strongly agree	N/A
We feel indebted to EPEA and/or MBDC for what they have done for us.	○	○	○	○	○	○
The company's employees share close social relations with the employees from EPEA and/or MBDC.	○	○	○	○	○	○
Our relationship with EPEA and/or MBDC can be defined as "mutually gratifying."	○	○	○	○	○	○
We expect that we will be working with EPEA and/or MBDC far into the future.	○	○	○	○	○	○
The collaboration with suppliers was extremely important to implement C2C standards.	○	○	○	○	○	○
The company extensively collaborated with scientific partners, e.g. universities, in order to implement C2C standards.	○	○	○	○	○	○
The company extensively exchanged knowledge with other companies in order to implement C2C standards.	○	○	○	○	○	○

Other (please specify)

Appendix A: Online survey

Please indicate your agreement on the following statements

	Strongly disagree	Disagree	Neither nor	Agree	Strongly agree	N/A
The company's **R&D skills** were more than adequate for implementing the C2C standards	○	○	○	○	○	○
The company's **engineering skills** were more than adequate for implementing the C2C standards	○	○	○	○	○	○
The company's **R&D resources** were more than adequate for implementing C2C standards	○	○	○	○	○	○
The company's **engineering resources** were more than adequate for implementing the C2C standards	○	○	○	○	○	○
In the course of implementing C2C standards, the company made fundamental changes to the existing product and processes	○	○	○	○	○	○
The innovation process had to be significantly adapted to the C2C standards	○	○	○	○	○	○
The implementation of C2C standards had a significant impact on the entire value chain of the company	○	○	○	○	○	○
The C2C certified products were mainly introduced as a new product line	○	○	○	○	○	○
The implementation of C2C standards is well incentivized by political regulations	○	○	○	○	○	○
The top management provided guidance during the implementation of C2C standards	○	○	○	○	○	○
The implementation of C2C standards would be easier if more companies from the industrial sector would implement it as well	○	○	○	○	○	○

Experiences during the implementation of C2C standards that you would like to emphasize

[   ]

Did the company's internal stakeholder (e.g. marketing, product development, sales) show resistance to the implementation of C2C standards?

○ Never   ○ Often
○ Rarely   ○ Always
○ Sometimes   ○ N/A

Comments:
[   ]

Appendix

In general, how would you agree or disagree to the following statements for your company (not only specific to C2C implementation)

	Strongly disagree	Disagree	Neither nor	Agree	Strongly agree	N/A
We easily replace one set of abilities with a different set of abilities to adopt a new technology	○	○	○	○	○	○
We tend to oppose new technologies that cause our manufacturing facilities to become obsolete	○	○	○	○	○	○
We are very willing to sacrifice sales of existing products in order to improve sales of our new products	○	○	○	○	○	○
We will not aggressively pursue a new technology that causes existing investments to lose value	○	○	○	○	○	○
We support projects even if they could potentially take away sales from existing products	○	○	○	○	○	○

After the implementation of C2C standards, the company's C2C certified product(s)…

	Strongly disagree	Disagree	Neither nor	Agree	Strongly agree	N/A
Overall, met or exceeded **sales** expectations	○	○	○	○	○	○
Met or exceeded **profit** expectations	○	○	○	○	○	○
Met or exceeded **return on investment** (ROI) expectations	○	○	○	○	○	○
Met or exceeded **market share** expectations	○	○	○	○	○	○
Met or exceeded **customer** expectations	○	○	○	○	○	○

Other (please specify)

Please indicate your agreement on the following statements

	Strongly disagree	Disagree	Neither nor	Agree	Strongly agree	N/A
We would recommend the implementation of C2C to our business partners	○	○	○	○	○	○
We are pleased to be associated with C2C	○	○	○	○	○	○
The implementation of C2C spurred innovation in the company	○	○	○	○	○	○
The company's image has improved because of the C2C implementation	○	○	○	○	○	○
The cost of C2C implementation is too high compared to the financial benefit	○	○	○	○	○	○
The cost of C2C certification is too high compared to the financial benefit	○	○	○	○	○	○

Other (please specify)

Please indicate your company's overall satisfaction with the C2C implementation

Very low	Low	Moderate	High	Very high	N/A
○	○	○	○	○	○

Other (please specify)

Appendix A: Online survey

Your functional department in the company

- ○ Innovation Management
- ○ Marketing
- ○ Production
- ○ PR / Communication
- ○ Other (please specify)
- ○ Research & Development
- ○ Sales
- ○ Sustainability
- ○ Executive Board

[   ]

Your position level

- ○ Employee
- ○ Lower management
- ○ Middle management
- ○ Top management

Number of years in the current position

- ○ 0 - 3 years
- ○ 4 - 7 years
- ○ 8 - 11 years
- ○ more than 11 years

Number of years with the company

- ○ 0 - 3 years
- ○ 4 - 7 years
- ○ 8 - 11 years
- ○ More than 11 years

Your gender

- ○ Female
- ○ Male

Your age

[⇅]

Appendix

The company's industry sector

- [ ] Building materials
- [ ] Chemicals
- [ ] Fast Moving Consumer Goods (FMCG)
- [ ] Floor covering
- [ ] Other (please specify)

- [ ] Furniture
- [ ] Home Care (cleaning supplies)
- [ ] Interior design
- [ ] Office supplies

- [ ] Packaging
- [ ] Paper
- [ ] Personal care
- [ ] Textile and fabric

The company's country of origin

[dropdown]

Commerce focus

- ( ) Business-to-business
- ( ) Business-to-customer
- ( ) Business-to-government
- ( ) Other (please specify)

Year of company foundation

- ( ) 1900 or earlier
- ( ) 1900 - 1950
- ( ) 1951 - 1980
- ( ) Other (please specify)

- ( ) 1981 - 2000
- ( ) 2000 - 2010
- ( ) 2011 or later

Company size (in number of employees)

- ( ) 1 - 10 employees
- ( ) 11 - 50 employees
- ( ) 51 - 100 employees
- ( ) 101 - 500 employees
- ( ) Other (please specify)

- ( ) 501 - 1.000 employees
- ( ) 1.001 - 5.000 employees
- ( ) 5.001 - 10.000 employees
- ( ) more than 10.000 employees

If you are interested in the overall research results, please enter your email-address (it will be handled confidential and independent of your survey results).

[　　　　　　　　　　　]

If you would be interested in a deeper case study analysis in the context of this research effort, please enter your email-address (it will be handled confidential and independent of your survey results).

[　　　　　　　　　　　]

Thank you very much for your time and contribution!

8.2 Appendix B: Descriptive results on responding companies

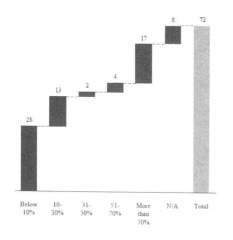
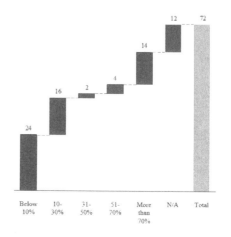

Figure 29: Share of C2C certified products of respondent companies

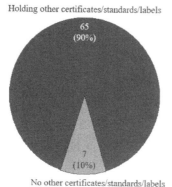

Most stated labels/certificates/standards

- ISO standards
- FSC – Forest Stewardship Council
- LEED – US Green Building Council
- Öko-Tex
- Energy Star

- In addition several national or industry-specific labels were mentioned

Figure 30: Other labels for environmental or social responsibility

Appendix B: Descriptive results on responding companies

The highest C2C certification level that one (or more) of your products achieved

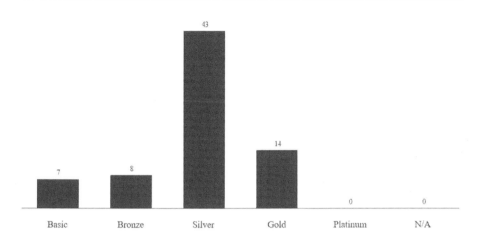

Figure 31: Highest certification level of C2C products of respondent companies

Please indicate the extent to which you believe that at this point in time C2C philosophy, standards, and methods have been implemented throughout your company

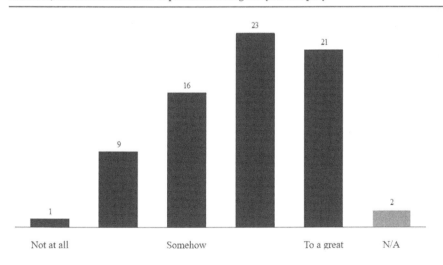

Figure 32: Extent of C2C implementation of respondent companies

Appendix

8.3 Appendix C: Inter-item correlation tables

		M9	M8	M4	M14	M12	M6	M11	M7	M1	M3
Factor 'competitive pressure'											
M9	The competition from other C2C certified companies	1									
M8	The potential loss of market share if the company does not implement C2C standards	.611	1								
M4	The expectation to contribute to the company's cost reduction	.533	.595	1							
M14	The expectation to better comply with current legislation	.387	.514	.483	1						
Factor 'expected benefits'											
M12	The expectation to improve customer satisfaction					1					
M6	The expectation to increase sales					.489	1				
M11	The expectation to be perceived as a market leader					.389	.375	1			
M7	The expectation of C2C to be a source of new opportunities					.447	.383	.399	1		
Factor 'strategic decision'											
M1	The fit of the C2C concept with the company's philosophy									1	
M3	The company CEO's / top management's strong endeavor to achieve the C2C certification									.528	1

Figure 33: Inter-item correlations for motivational factors

Appendix C: Inter-item correlation tables

	PAR1	PAR3	PAR4	PAR2	IMP2	IMP7	IMP1	IMP4	TS7	TS5	TS6
Factor 'relationship with certification partner'											
PAR1 We feel indebted to EPEA and/or MBDC for what they have done for us	1										
PAR3 Our relationship with EPEA and/or MBDC can be defined as "mutually gratifying"	.626	1									
PAR4 We expect that we will be working with EPEA and/or MBDC far into the future	.456	.615	1								
PAR2 The company's employees share close social relations with the employees from EPEA and/or MBDC	.510	.443	.345	1							
Factor 'C2C anchorage'											
IMP2 The company has identified specific persons and positions responsible for C2C implementation					1						
IMP7 The extent to which you believe that at this point in time C2C philosophy, standards, and methods have been implemented throughout your company					.351	1					
IMP1 The company has integrated C2C in procedures and work instructions					.399	.583	1				
Factor 'integration into supply and value chain'											
IMP4 The company has obliged its supply base to supply according to C2C								1			
TS7 C2C implementation had a significant impact on the entire value chain of the company								.624	1		
Factor 'degree of change'											
TS5 In the course of implementing C2C, the company made fundamental changes to the existing product and processes										1	
TS6 The innovation process had to be significantly adapted to C2C										.610	1

Figure 34: Inter-item correlations for organizational context factors